# Invertebrate Animals

# Collection and Preservation

# Invertebrate Animals

## Collection and Preservation

Compiled by
Roger J Lincoln & J Gordon Sheals

British Museum (Natural History)
Cambridge University Press

·

Published by the British Museum (Natural History), London
and the Syndics of the Cambridge University Press
The Pitt Building, Trumpington Street, Cambridge CB2 1RP
Bentley House, 200 Euston Road, London, NW1 2DB
32 East 57th Street, New York, NY 10022, USA
296 Beaconsfield Parade, Middle Park, Melbourne 3206, Australia

First Published 1979

Printed in Great Britain by Butler & Tanner Ltd,
Frome and London

ISBN 0 521 22851 4 hard covers
ISBN 0 521 29677 3 paperback

**Library of Congress Cataloguing in Publication Data**

Invertebrate animals.
    Includes bibliographical references and index.
    1. Invertebrates – Collection and preservation.
I. Lincoln, Roger J. II. Sheals, John Gordon, joint
author.
QL362.8.I578 1979      579      79-14530

# Contents

# Introduction

This manual is intended to replace handbook No 9A of the series *Instructions for Collectors* entitled *Invertebrate Animals other than Insects*. With the exception of the Mollusca, and possibly the spiders, the 'non-insectan' invertebrates have not attracted a great deal of amateur attention, and a good part of the museum systematists' requirements in these groups is now met by well-equipped expeditions with scientifically trained participants. For this reason it was felt that handbook 9A might be usefully replaced by a more detailed treatment of collecting and preservation techniques. At the same time it is hoped that the new approach will not deter amateur naturalists. An attempt has been made to explain the underlying principles of fixation and preservation, and the expanded biological and distributional notes may be helpful to collectors with expert knowledge of only a few groups.

While these notes are intended primarily for persons collecting on behalf of the British Museum (Natural History), it is hoped that they will be found useful to others, particularly students on field courses. The Museum always has a need for material of some kind from all parts of the world, and when planning expeditions collectors are invited to consult with the scientific staff, who are always willing to advise on the sort of work that is likely to be profitable in a particular region. In selected cases collectors may be supplied with preservation materials and lent a certain amount of equipment. Preliminary enquiries should be addressed to: The Keeper of Zoology, British Museum (Natural History), Cromwell Road, London SW7 5BD.

It is most important that collectors should familiarize themselves with, and scrupulously observe, regulations and legislation relating to conservation and to the export and import of specimens. Collecting can only be supported when there is a real educational or scientific need, and casual collecting with inadequate documentation can never be justified. In most cases, careful collecting of the groups dealt with in this handbook is unlikely to cause any long-term damage to natural populations since these animals are usually very numerous, small in size, difficult to isolate from their habitats, and have relatively short generation times. Much more at risk from indiscriminate field work is the habitat itself, which can be seriously affected by excessive disturbance or damage, often resulting in widespread injury to the natural flora and fauna.

Whilst we accept full responsibility for any errors in the text we have drawn heavily on the expertise of our colleagues in compiling the account, and thanks are due to the following for information and advice: Dr C. R. Curds (Protozoa); Dr P. F. S. Cornelius (Coelenterata and Ctenophora); Mr S. Prudhoe (Platyhelminthes, Mesozoa, Nemertinea, Aschelminthes and Acanthocephala); Miss P. L. Cook (Entoprocta and Bryozoa); Mr E. F. Owen (Brachiopoda); Mr J. F. Peake (Mollusca); Mr R. W. Sims (Sipuncula, Echiura, Priapulida and Oligochaeta); Dr J. D. George (Porifera, Polychaeta and Pogonophora); and Miss A. M. Clark (Echinodermata, Hemichordata, Urochordata and Cephalochordata). Thanks are also due to Mr R. H. Harris for advice on general matters of fixation and preservation and to Mrs S. Chambers for preparing the plates for Chapter 1.

*September, 1976*                                    R.J.L. & J.G.S.

Chapter 1

# The Main Invertebrate Groups

The division of the animal kingdom into vertebrates and invertebrates is a common practice in text books and educational courses in zoology, but this separation is quite arbitrary. Of the twenty or so phyla that make up the animal world all except one, and that only a sub-phylum, are invertebrate, representing at least 95 per cent of all known animal species. In this chapter each of the major invertebrate phyla is treated in turn to provide a brief description of the morphology of the organisms, a further classification of the group, its habitat range, and useful methods of collection and preservation. In some cases the collection and preservation of a particular group of organisms is a straightforward procedure, but some require special techniques for collection or extraction and special treatment for effective preservation. Each group is illustrated with one or more figures to help distinguish its members and to show something of the diversity of form within that group.


## Protozoa (Plate 1a–j)

It is impossible to define this heterogeneous group of microscopic animals in terms of a small number of characters. They are sometimes described as being unicellular but this is misleading since the single cell of the protozoan is not really the equivalent of a single cell of a multicellular animal. The term acellular has also been used but this adjective is perhaps just as misleading. Although the majority of Protozoa occur as solitary individuals there are a number of colonial forms.

The following outline classification of the Protozoa is that used by Mackinnon & Hawes (1961). Although this classification may now be considered inadequate by some taxonomists, and alternative schemes have been proposed more recently (Honigberg et al., 1964), it is satisfactory for present purposes as a simple scheme with which to illustrate the range of form within the group. The phylum is divided into four classes: Rhizopoda, Mastigophora, Sporozoa and Ciliata.

The Rhizopoda are typified by the presence of pseudopodia. These are temporary extensions of the body that function as organs of locomotion and feeding. The form of the pseudopodia varies considerably – they may be blunt rounded processes (lobopodia) as in the familiar

*Amoeba*, or long slender filamentous structures (filopodia) that may branch and anastomose (rhizopodia), and in some species the pseudopodia are stiffened by axial tubules (axopodia). The diversity of the rhizopod protozoans is reflected in the sub-divisions of the class. The order Amoebina comprises naked organisms that normally have blunt lobopodia, the Testacea and Foraminifera have a firm outer shell or test and have either filopod or rhizopod pseudopodia, the Heliozoa lack an outer test but have their pseudopodia stiffened by axial rods (axopodia), and the Radiolaria have an internal siliceous capsule providing skeletal support to the body and have fine filamentous pseudopodia.

The test may be a loose structure formed by the adherence of organic or inorganic particles to the surface of the organisms as in some Testacea, or it may be a solid shell formed of organic or inorganic material which imposes a rigid shape on the animal. The calcareous tests of some Foraminifera and the siliceous shells of some Radiolaria are of great architectural complexity and beauty.

The Mastigophora comprises those protozoans often referred to as 'flagellates'. They all possess one or more flagella – long whip-like structures – that are used mainly for locomotion but also assist in feeding. The group is divided into two subclasses, the Phytomastigina and Zoomastigina. The Phytomastigina are characterized by the presence of the photosynthetic pigment chlorophyll, and usually have only one or two flagella. The Zoomastigina on the other hand do not have chlorophyll and thus cannot synthesise organic compounds, but instead have to rely on food nutrients for the intake of these substances. Further, the zoomastigines may have many flagella some of which may form more complex organelles. Among the familiar phytomastigines are the solitary *Euglena* and *Chlamydomonas*, the colonial *Volvox* and the dinoflagellates *Ceratium* and *Noctiluca*. The medically important pathogens of the genus *Trypanosoma* belong to the Zoomastigina.

Flagellate protozoa are found in all types of habitat – in the sea, in fresh water and in the soil. They may be free-living, solitary or colonial, and there are parasitic forms as well.

The Sporozoa are exclusively entozoic, i.e. they all live within the body cavity or tissues of an animal host. The members of the group have complex life histories; the group deriving its name from the

Plate 1 Free-living Protozoa
a, *Euglena*, a phytoflagellate; b, *Ceratium*, a dinoflagellate; c, a spherical radiolarian; d, a non-spherical radiolarian; e, a colonial flagellate; f, *Difflugia*, a testate amoeba; g, *Carchesium*, a colonial ciliate; h, *Tetrahymena*, a solitary ciliate; i, *Tintinnopsis*, a loricate ciliate; j, *Discorbis*, a foraminiferan.

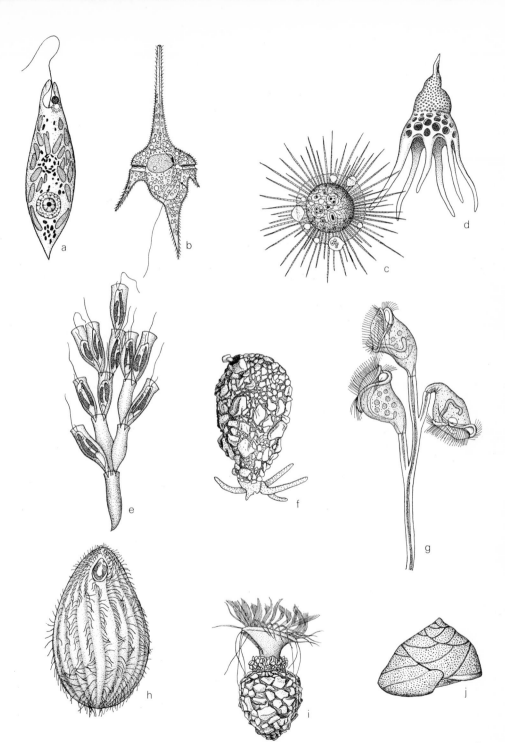

3

process of spore formation (sporogony) that forms part of the life-cycle. In most sporozoans the life-cycle includes an intracellular stage, and the group contains a number of medically important species.

The Ciliata or 'ciliates' are, as the name suggests, typified by the presence of cilia on the surface of the body. The condition is illustrated clearly in the familiar *Paramecium* and *Tetrahymena* where the body is covered with a 'coat of fine hairs'. The body shape within the group is very variable, as is the arrangement of the cilia which may be modified into complex organelles. As with the flagellates, the ciliates are very common, occurring in all types of habitat. Many are free-living, but there are stalked attached forms as well as a host of parasitic species.

**Collecting**

The majority of Protozoa require expert techniques on the part of the collector, and here it is intended to deal only with certain marine forms that either have hard shells or have hard skeletal structures (Foraminifera and Radiolaria). These organisms are comparatively easy to collect. Good accounts of the techniques of protozoology are given by Wenyon (1926), Kirby (1950) and Mackinnon & Hawes (1961).

Living Foraminifera may be collected from corallines and other seaweeds either by picking them off under a lens or by the use of a coarse sieve with bolting-silk fastened beneath it. The sieve should be immersed in sea water and handfuls of seaweed, etc., shaken over it. The bolting cloth will catch the animals that pass through the sieve. Mud and fine sand from the sea floor may be put into vessels of sea water and stirred. The living foraminiferans will sink to the bottom and the lighter particles can be poured off with the water. Pelagic species can be collected with a tow-net (p. 110).

Large areas of the ocean floor are rich in the calcareous shells of dead Foraminifera. The Globigerina Ooze, which covers nearly one hundred and thirty million square kilometres (fifty million square miles) of the ocean floor, consists almost entirely of them. Dredgings from such areas can be dried and sifted, and the finer siftings placed in a bowl of water and stirred. The more delicate shells will float and may be skimmed off, while the heavier shells sink and may be dried after the water has been removed.

Many species of Radiolaria live at or near the surface of the sea, where they sometimes occur in great numbers. They may be collected with a tow-net or in a bucket of sea water. The siliceous shells of dead radiolarians sink to the bottom and form the Radiolarian Ooze, which covers an area of more than five million square kilometres (two million square miles). This deposit is typically found in very deep water in the

4

tropical regions of the Pacific and Indian Oceans. They can be collected, like those of the Foraminifera, by dredging, and may be dried in a similar way.

**Preserving**
Living animals should be placed first in 50 per cent alcohol and then in 70–90 per cent alcohol. Alternatively, they may be placed directly in 3–5 per cent neutral formalin solution. If material is being collected for subsequent cytological examination specialist literature on fixation techniques should be consulted.

*References*
Honigberg, B. M., Balamuth, W., Bovee, E. C., Corliss, J. O., Gojdics, M., Hall, R. P., Kudo, R. R., Levine, N. D., Loeblich, A. R., Weiser, J., & Wenrich, D. H. 1964. A revised classification of the Phylum Protozoa. *J. Protozool.* **11** (1) 7–20.
Kirby, H. 1950. *Materials and Methods in the Study of Protozoa.* Berkeley and Los Angeles: University of California Press, 72 pp.
Mackinnon, D. L. & Hawes, R. S. J. 1961. *An Introduction to the Study of Protozoa.* Oxford: Clarendon Press. 506 pp.
Wenyon, C. M. 1926. *Protozoology*, vol. 2. London: Bailliere, Tindall & Cox. pp. 1291–1335.

# Porifera (*Sponges*) (Plate 2a–h)

Sponges are the most primitive multicellular animals for, although some of their component cells are specialized to perform certain tasks, they all display a considerable amount of independence and properly differentiated tissues and organs are not present.

All sponges live in water, spending their life attached to rocks, shells, weeds and other suitable surfaces. They are extremely variable in shape, size, colour and consistency and it is not possible to set out a brief comprehensive description by which all can be positively recognized. The bath sponge is a familiar object, but sponges can also grow as irregular masses or thin encrustations. Deep-water sponges tend to have more regular shapes than those found in shallow waters, and they can resemble vases, fans, tubes, plants, etc. Sponges range in size from a few millimetres to several metres, they can be brilliantly coloured, and can have a consistency varying from soft and 'spongy' to hard and stone-like. Owing to their sedentary way of life they are often mistaken for seaweeds or groups such as coelenterates, Bryozoa and tunicates.

Sometimes their characteristic spongy consistency and porous surface helps to distinguish them, but the only reliable way of identifying them is to make a microscopical examination of their internal structure.

Although there are a number of freshwater species living in rivers, lakes, ponds and reservoirs, attached to the stems of water plants, pebbles, piles of bridges, etc., the vast majority of sponges are marine. They abound in all seas wherever there are submerged rocks, timbers, shells, corals and so on to form a suitable substratum. Most sponges prefer shallow waters, although some groups such as the 'glass sponges' are found at great depths. Many species have become adapted to intertidal conditions and these live in sheltered situations up to mid-tide level. They occur on various rock surfaces, cave roofs, seaweeds (especially *Laminaria* holdfasts) and they are often found on crabs and other animals. Some sponges make borings in shells or calcareous rocks.

**Collecting**
Intertidal and shallow-water species can be easily collected without special equipment and species living in deeper waters can be taken by diving or dredging. Wherever possible whole sponges should be collected still attached to their immediate substratum, and this applies particularly to the boring forms. A hammer and chisel may be necessary, but if this method of collecting is impracticable the sponge can be gently prised from the surface with a knife. If a sponge is too large to preserve whole, a portion containing all the typical features may be cut off and kept. Specimens should be kept apart to prevent an exchange of surface spicules – these skeletal elements are used in identification. A convenient way of doing this is to use polythene bags. Each specimen as it is collected, should be put, together with its specimen label, in a separate bag and preservation fluid immediately added. Delicate specimens should be put in rigid tubes or jars, or padded with tissue paper (**not** cotton wool) if bags have to be used. Bags containing specimens should be tied up securely, and to safeguard against leaks these bags can be stored in larger bags of heavy-duty polythene. Useful study series

Plate 2 Porifera
a, *Euplectella* ('Venus's Flower Basket'), a deep-sea 'glass' sponge, length 210 mm; b, *Platychalina*, a branching demosponge, length 290 mm; c, *Grantia*, the 'purse sponge', length 40 mm; d, *Leucosolenia*, an Australian calcareous sponge, length 30 mm; e, *Esperiopsis*, a deep-sea demosponge, length 210 mm; f, *Haliclona*, a branching demosponge, length 170 mm; g, *Axinella*, a vase-shaped demosponge, length 70 mm; h, *Spinosella*, a tubular-shaped tropical demosponge, length 300 mm.

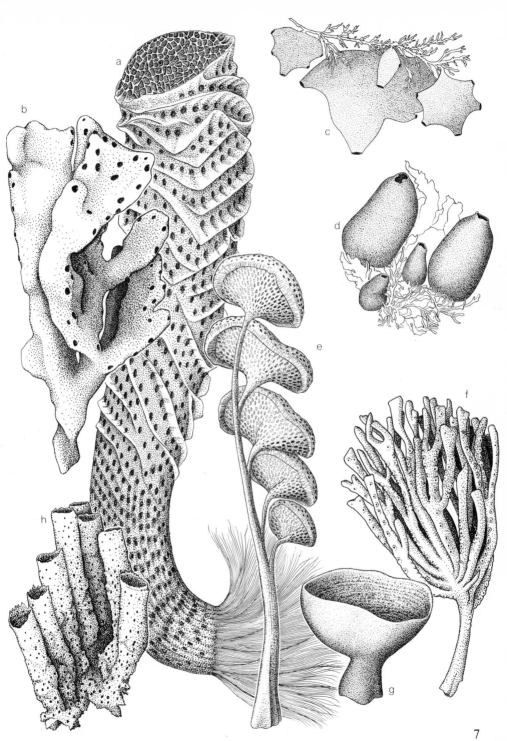

7

of marine species can be made by collecting two specimens of each sponge, keeping one dry and preserving the other in alcohol. It is not necessary to dry freshwater sponges. When sampling an area, great care should be taken to ensure that a species is not collected to extinction. Except in very special cases, small sponges less than 2 mm in diameter should not be collected since they are of limited taxonomic value.

In addition to basic collection data (p. 138) notes should be made in a separate notebook on the shape, size (overall measurements if only a portion of the sponge is collected), colour and consistency of the living animal. Records should also be made of any unusual changes that occur on preservation such as an excessive discharge of mucus or heavy extrusion of spicules. If additional notes can be made, observations on the frequency of the sponge in the area, a list of biological associates (including commensals and predators), and an expanded description of the habitat could all be useful. This should include notes on the type of substratum; position of the sponge on the substratum (e.g. on top of or below boulders, in crevices, on vertical or horizontal rock faces, on cave roof, partially buried in estuarine mud, etc.); relation of the sponge to local conditions (e.g. exposed to wave-action or swirl, protected by a heavy canopy of weed, etc.) and exposure to light. Physical and chemical records are always useful (e.g. water temperature, salinity and light readings). Good colour transparencies of the living animal and the habitat are invaluable.

### Preserving
Sponges should be preserved as soon as they are collected since they deteriorate quickly. Alcohol is the most useful preservative. Specimens should be completely immersed in 50 per cent alcohol, and after about 12 hours transferred to clean spirit of the same strength. After a further 12 hours they should be transferred to 70 per cent alcohol. Warm Bouin's fluid (p. 128) is recommended when fixed material is required for histological studies. Formalin solution is not recommended as a preservative for sponges.

Marine sponges can be allowed to dry in a well-ventilated, cool place, but before drying they should be soaked for two hours in fresh water to remove salts.


## Coelenterata (Plate 3a, c–f)

This phylum includes the hydroids, jellyfish, sea anemones, corals and several smaller groups, and although immensely variable in appearance

8

all these animals have a similar fundamental organization. They are radially symmetrical and the body wall consists of three basic layers – an outer epidermis, an inner layer of cells lining the digestive cavity and between them a gelatinous layer with rather fewer cells called the mesogloea. The centrally situated mouth is almost always surrounded by one or more circles of tentacles that aid in the capture and ingestion of food. The tentacles are equipped with characteristic sting cells – nematocysts – that are used to trap and paralyse the prey. Although coelenterates are basically tentaculate and radially symmetrical two distinct structural types are found within the phylum – a form known as a polyp that is almost invariably sessile and a form known as a medusa that is nearly always free-swimming. The polyp consists essentially of a cylindrical body with the oral end, bearing the mouth and tentacles, directed upwards. The medusa on the other hand is shaped like an umbrella with the mouth and surrounding tentacles located on the concave undersurface. Some coelenterates exist only as polyps, some only as medusae, while others pass through both stages during their life-cycle. The phylum is made up of four classes: Hydrozoa, Scyphozoa, Cubozoa and Anthozoa.

## Hydrozoa

This class includes a number of very common species and much of the marine growth that is found on shells, rocks and wharf pilings is formed by hydrozoan coelenterates. Most consist of sessile, branching polyp colonies, many of which give rise to small free-swimming medusae. However, in some the polypoid phase is much reduced, and members of one order – the Siphonophora – which includes the notorious *Physalia physalis* (Portuguese man-of-war), exist as large pelagic colonies comprising modified polypoid and medusoid individuals. In *Physalia* the extremely long tentacles (dactylozooids), which hang from the colony rather like a drift net, are covered with nematocysts which can inflict extremely painful stings. Although the majority of the Hydrozoa are marine, some (including the familiar 'hydras' and a few species of medusae) live in fresh water.

### Collecting
Many sessile marine species in the littoral zone and in shallow sublittoral waters can be collected quite easily without special equipment, although a hammer and cold chisel may be useful for chipping off portions of rock bearing hydroid growths. Sessile forms in deeper water

can be obtained by dredging, although, as with other marine groups, forms living on hard substrates in moderately deep water are best collected by diving.

The free-swimming medusae can be caught by tow-netting, which can be more successful at night when the medusae tend to be more abundant at the surface, and the larger pelagic forms such as Siphonophora can be taken with dip-nets or with a bucket.

Species living in fresh and brackish water can be obtained by similar methods, although hydras are best collected by examining small samples of aquatic plants in a dish of water under a lens or binocular microscope. These animals are often found clinging to the undersides of leaves and can be removed with a small pipette.

**Preserving**

With the possible exception of some Siphonophora, the Hydrozoa should be anaesthetized before being fixed, and the most generally useful anaesthetic substance is menthol. The animals should be allowed to extend in a vessel of water (sea water, brackish or fresh water as appropriate) and a few crystals of menthol scattered on the surface. The time required for anaesthetization varies, but the animals can be judged to be sufficiently relaxed when they cease to respond to probing. Alternative anaesthetic substances are MS 222 (p. 124), magnesium sulphate, magnesium chloride, propylene phenoxetol and stovaine (p. 126). Propylene phenoxetol is particularly good, and fast-acting, for hydroids and small medusae, provided it is added slowly to avoid mechanical disturbance and contraction of the tentacles. If other substances are not available, very dilute formalin solution can be used as an anaesthetic for some hydroids and small medusae, while some medusae can be successfully relaxed by allowing them to warm up slowly to room temperature.

For general taxonomic purposes Hydrozoa can be fixed in 20 per cent buffered formalin solution. The material can be preserved in 10 per cent formalin solution or 70 per cent alcohol.

## Scyphozoa

This class includes the larger jellyfish. The polypoid phase is usually either completely suppressed or restricted to a developmental stage of comparatively short duration. The group is exclusively marine, and whilst some orders have a rather restricted distribution, the group as a whole occurs in all the oceans of the world. Whilst some are exclusively deep-water forms, the majority of Scyphozoa inhabit coastal waters, and with their stinging nematocysts they are often a nuisance to swimmers. One order, the Stauromedusae, has adopted a sessile existence with some species occurring on intertidal algae.

### Collecting
The smaller pelagic species can be collected by tow-netting and the larger jellyfish can be taken with a dip-net or with a bucket. Care should be taken to avoid contact with the stinging tentacles of the larger forms. Stauromedusae can be collected with a portion of the algal substrate or if attached to a rock can be removed using a blunt knife.

### Preserving
Pelagic Scyphozoa can be fixed in 20 per cent buffered formalin solution and preserved in 10 per cent formalin solution. Anaesthetization is not required in the planktonic species. Stauromedusae can be relaxed in menthol, and MS 222 is effective with the British species. Fixation of Stauromedusae is best done in neutral 10 per cent formalin solution which is also adequate for storage, or the specimens can be transferred to 70 per cent alcohol through gradual steps.

## Cubozoa

This group comprises a single order, the Cubomedusae, which has recently been given class status because of its peculiar life-history in which a polyp develops into a single medusa. All known species are marine. The stings of certain large Australian species are extremely virulent, fast-acting and frequently fatal. As yet there is no fully effective antidote.

### Collecting and preserving
General methods as for Scyphozoa. Particular care should be taken with the venomous species which should not be allowed to touch the skin.

# Anthozoa

The Anthozoa are solitary or colonial marine coelenterates in which the medusoid phase is completely suppressed. This class is made up of two subclasses: Zoantharia and Octocorallia. In the Zoantharia, which includes the sea anemones (order Actiniaria) and the true or stony corals (order Scleractinia), the tentacles are numerous, rarely branched and usually arranged in multiples of six. By contrast, in the Octocorallia, which includes such familiar forms as sea pens and sea pansies (order Pennatulacea), sea fans and horny corals (order Gorgonacea) and soft corals (order Alcyonacea), the tentacles are eight or multiples of eight in number and pinnate – that is with side branches rather like a feather.

The anemones are solitary polyps, virtually devoid of any skeletal structure. They vary considerably in size, but while some tropical species may measure 1 metre or more across the oral disc, the majority range in diameter from about 10 to 40 mm. Many species are very brightly coloured and their tentacles and oral disc frequently display quite spectacular colour patterns. Actiniaria are found in all seas but they reach their greatest abundance and diversity in warmer waters. Some species have been found at considerable depths, but typically they are found in the shallow sublittoral and on rocky shores where large numbers of individuals of the same species can often be found clustered together in rock crevices. Many species burrow in sand, and some are commensal with hermit crabs, living attached to the gastropod shells which the crabs inhabit.

The polyps of true corals resemble small anemones but they are usually colonial and produce an external skeleton of calcium carbonate. True corals are found in all the oceans of the world, but again the richest faunas are found in warmer seas. The reef-building species require warm shallow waters and as a result are found only along certain continental and island shores in tropical and sub-tropical regions.

The Octocorallia are all colonial, and the colonies are supported by a calcareous or horny skeleton. They vary considerably in form and the common names of the various groups are often a reflection of their particular growth patterns. Although some species extend into temperate or even polar regions, the Octocorallia are most abundant in the

Plate 3 Coelenterata and Ctenophora
a, *Rhizostoma*, a large jellyfish; b, *Pleurobrachia*, a ctenophore or comb-jelly; c, *Campanularia*, a thecate hydroid; d, *Pennatula*, a sea pen; e, *Oculina*, a branched coral; f, *Actinothoe*, a sea anemone.

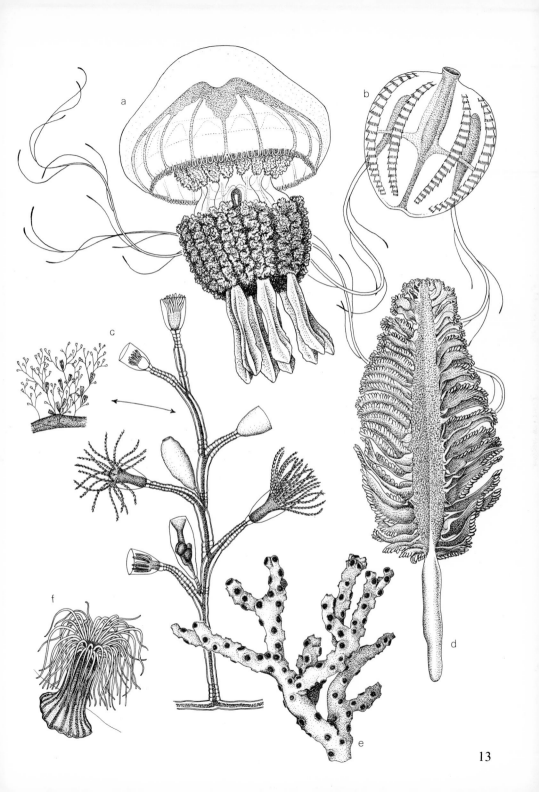

13

warmer parts of the world and they attain their greatest diversity in the Indo-Pacific ocean. They occur mainly in shallow coastal waters but some sea fans and soft corals have been recorded from depths exceeding 3000 m.

### Collecting
Species forming large colonies are best collected by hand, with the aid of hammer and chisel where appropriate. Dredging and similar techniques can also be employed but frequently result in damage to the specimens, particularly in the case of much-branched forms. Solitary species are often collected attached to a portion of rocky substrate from which they can be removed if necessary with a blunt scalpel, or in the case of the larger anemones by a thin wedge driven carefully under the basal disc. Burrowing anemones should be dug out carefully with a small spade.

### Preserving
Forms having a calcareous skeleton should be preserved in 70 per cent alcohol since the calcareous material is dissolved by formalin and it is not certain that even neutralized formalin will leave the finer structures unharmed. If only the skeleton is required, as for example with true corals, the specimen can be placed for several days in a 50/50 solution of household bleach and water and then rinsed and dried; or simply sun-dried, and then bleached at a later date. Gorgonians and antipatharians need only be sun-dried. Other Anthozoa can be fixed in 20 per cent buffered formalin and preserved either in a 10 per cent solution or in 70 per cent alcohol. Menthol is a universal but slow-acting anaesthetic. MS 222 works well with many species, but not for example with the snakelocks anemone, *Anemonia sulcata*. It is probably advisable to try a variety of relaxing agents for each species, in the order menthol, MS 222, magnesium chloride and magnesium sulphate.

## Ctenophora (*Comb jellies, sea gooseberries*) (Plate 3b)

This small phylum of exclusively marine animals is made up of about 80 species. Their bodies are gelatinous and, typically, subspherical, ranging in diameter from about 3 to 50 mm. However, some species are elongated as in the case of the well-known Venus girdle, *Cestum veneris*, which has the form of a flattened gelatinous band. This animal may attain a length of over 1·5 m. Ctenophores are usually transparent, although structures such as tentacles and the ciliated bands known as comb-rows may be coloured white, orange or purple. Like the coelenterates, ctenophores are radially symmetrical and in general plan their bodies resemble those of coelenterate medusae. However, there are certain important differences – for example, delicate muscle cells are found in the intermediate gelatinous layer of the body of ctenophores, their digestive system is much more elaborate, and with few exceptions they lack nematocysts.

Apart from four aberrant genera classified in the order Platyctenea that have taken up a creeping, or in the case of the genus *Gastrodes* a parasitic, existence, ctenophores are pelagic animals. They are feeble swimmers and are frequently concentrated by winds and currents into enormous aggregations. After storms their bodies are often found strewn in masses along beaches. The group as a whole has a worldwide distribution although the range of some species is limited.

### Collecting

Planktonic ctenophores are taken by tow-net or dip-net and it is often an advantage to collect the animals at night since at this time they are more abundant near the surface. Creeping forms may be found on certain corals, and species of *Coeloplana* have been found on alcyonarians in the Indian and western Pacific oceans.

### Preserving

Ctenophores, particularly the larger species, are difficult to preserve. It is generally an advantage to anaesthetize the animals before fixing and this can be done by adding chloral hydrate crystals to the sea water containing them. The best general fixative for these animals is a chromic/osmic acid mixture (p. 130). The anaesthetized animals should be transferred into this mixture from the sea water, and left for 15 minutes to an hour according to size. Another useful general fixative is Flemming's solution (p. 130). A chromic/acetic acid fixative solution (p. 131) is recommended for *Cestum veneris*. These animals should be placed in a small quantity of sea water and a relatively large quantity of the fixative added. Material should be left in this solution for about 15 minutes.

After fixation ctenophores should be graded through alcohol to 70 per cent, i.e. they should be taken very slowly through 30, 40, 50 and 60 per cent alcohol to the 70 per cent spirit. All species of Ctenophora should be stored in 70 per cent alcohol, and formalin solutions should never be used. During fixation and subsequent grading in alcohol specimens of the larger atentaculate species, such as certain species of *Beroe*, which have a very wide mouth and pharynx, can be supported by placing the open end of a glass tube of suitable diameter into the mouth. After two or three days in 70 per cent alcohol the tube can be withdrawn.

## Platyhelminthes (*Flatworms*) (Plate 4b–i)

This phylum is made up of three classes of flattened soft-bodied animals that are constructed on a higher level of organization than the sponges and coelenterates. They have, for example, an excretory system and their reproductive organs are extremely complex. One of the classes – the Turbellaria – is composed mainly of free-living animals, but the other two classes, Trematoda and Cestoda, are entirely parasitic.

### Turbellaria (*Planarians*)

The body shape in the Turbellaria varies from ovoid to elongate and a head-like projection may be present. Most species are greyish, blackish or brownish, but a few are more brightly coloured and some may be green owing to the presence of symbiotic algae. Turbellarians are primarily aquatic and the majority are marine. There are, however, several freshwater species and a few are terrestrial, although the latter are found only in damp habitats such as moss and below dead wood on the forest floor. Although there are some pelagic species, the majority of the aquatic planarians are bottom-living forms and are found mainly in sand, mud and under stones and shells. Marine forms are common in the intertidal region and in shallow coastal waters.

**Collecting**
Planarians can be picked up with the fingers, with forceps or with a small paint brush and many freshwater species can be obtained by sweeping water plants with a fine-meshed net. Bottom-living forms in deeper water can be taken by dredging. In streams they can be caught by baiting with a small piece of raw meat.

Plate 4 Platyhelminthes and Nemertinea
a, *Lineus*, a heteronemertine worm; b, *Thyzanozoon*, a polyclad turbellarian; c, *Schistosoma*, a digenetic fluke or trematode; d, *Dendrocoelum*, a triclad turbellarian; e, *Diclidophora*, a monogenetic fluke; f, *Gyrocotyle*, a cestodarian; g, a tapeworm; h, scolex of a tapeworm, *Hymenolepis*; i, *Fasciola*, a digenetic fluke or trematode.

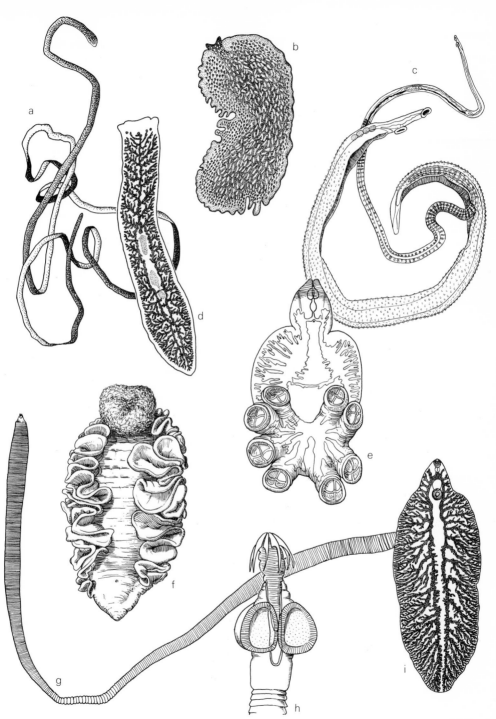

17

### Preserving

Turbellaria are preserved in 70–90 per cent alcohol. It is possible to place fresh specimens directly into the preservative, although it is generally advisable to subject them to a rapid fixation process before preservation. Hot alcohol* can be used as a rapid fixative, although this is not always a convenient method. The animals should first be placed in a dish of clean water (sea water for marine species). After a few minutes nearly all of the water is gently poured off and as the animals are extending the fixative fluid is poured over them. Alternatively, the animals can be dropped momentarily into hot water to kill them (usually in an extended condition) and then placed directly into 70–90 per cent alcohol for fixation and preservation. Planarians can also be fixed with 3–5 per cent formalin solution, but the results are generally less satisfactory.

## Trematoda (*Flukes*)

The trematodes or flukes are flattened ovoid unsegmented parasites, and as adults they occur almost invariably in vertebrate hosts. Some species, however, have a complex life-cycle and immature stages may be found in vertebrate and invertebrate animals. Adult flukes have one or more suckers that act as holdfasts and some species are also provided with hooks. As internal parasites the adults are found in a wide range of animals, but as external parasites they occur only on aquatic hosts.

### Collecting

Ectoparasitic flukes can be picked off the host with forceps. They are often found in the mouth cavity of the host and, in the case of fish, they are also found commonly on the fins and gills. Methods for collecting internal parasites are outlined on p. 119.

### Preserving

Living flukes should first be cleaned by thoroughly shaking them in a 1 per cent salt solution, and for most purposes they can be fixed in 10 per cent formalin solution and then stored in a 3–5 per cent formalin solution. After vigorous shaking in a 1 per cent salt solution most of

---

* The vapour given off by heated alcohol is highly flammable. The danger of fire will be lessened if the vessel used for heating the spirit is not too shallow. Its width need only be sufficient to allow the specimens to straighten out. Enamelled vessels are the most suitable.

the liquid should be poured off and replaced by a small quantity of clean solution. The shaking is continued and an equal volume of 10 per cent formalin solution is added quickly. The shaking should then be resumed at once – this is important because it prevents contraction in the specimens. When the flukes are rigid they should be transferred to the preservative (3–5 per cent formalin solution).

For fine anatomical work a saturated solution of corrosive sublimate (mercuric chloride, p. 131) is a better fixative. This should be used in the manner already described for the 10 per cent formalin. Flukes should be fixed in the saturated solution for a few minutes to an hour, depending on their size. The fixative is removed by washing with several changes of water (or 70 per cent alcohol) or by treatment with iodized alcohol (p. 131), and the material can be finally stored in 70–90 per cent alcohol. Bouin's or Zenker's fluids (pp. 128, 130) may also be used for fixation.

## Cestoda (*Tapeworms*)

Adult tapeworms have a rounded head that bears the organs of attachment (usually suckers and hooks), and in most cases the flattened body is elongate and made up of a string of similar sections called proglottides. As adults they occur almost exclusively in the intestines of vertebrates, but many species have complex life-cycles and immature stages may be found in both vertebrate and invertebrate hosts.

### Collecting

The methods of searching for intestinal parasites are outlined on p. 119. When removing tapeworms from the gut, care must be taken to avoid leaving their heads attached to the gut lining. Small specimens can be removed by scraping the gut membrane gently with the back of a knife. If the heads of comparatively large specimens are very firmly attached, gentle probing around the attached site with needles, in a dish of salt solution or water, may induce the worms to loosen their hold. Worms may detach themselves spontaneously if left soaking for a while in salt solution.

### Preserving

Tapeworms should be cleaned in a 1 per cent salt solution while still alive. They can either be carefully shaken in a jar nearly full of salt solution or washed in a shallow dish. The longer worms are apt to become tangled and to prevent this they should be rinsed separately, transferred with as little fluid as possible to a glass plate, and spread out.

Tapeworms must be fixed before preservation. After washing the worms should be kept as straight and as extended as possible. This can be done by laying them out, slightly stretched, on a sheet of glass and pouring a small quantity of the fixative solution over them. Alternatively, the fixative can be applied with a camel-hair brush. A better method is to allow the worm to hang head downwards so that it is stretched by its own weight. It should then be dipped quickly, a number of times, into a jar of the fixing reagent. Small forms may be drawn along the side of the vessel, after each dipping, so as to exert a slight pull on them. When the worms are rigid they should be left in a dish of the fixative for ten minutes to an hour, according to their size.

Formalin solution (5–10 per cent) may be used for fixing tapeworms. It should be used cold, and the specimens should afterwards be stored in a fresh 3–5 per cent formalin solution.

A better fixative is a modification of Schaudinn's solution, made up of roughly equal parts of a saturated solution of corrosive sublimate in water (see p. 132) and of 70 per cent alcohol, to which, preferably, a few drops of acetic acid should be added before use. This mixture should be used cold. Specimens treated with this fixative should be washed for several hours in water (running or frequently changed) or in several changes of 70 per cent alcohol. A better way of removing the corrosive sublimate is to treat them with iodized alcohol (p. 131). After the fixative has been thoroughly washed out the specimens can be preserved in 70–90 per cent alcohol or in 3–5 per cent formalin solution. Bouin's or Zenker's fluid can also be used for fixing tapeworms (see pp. 128, 130).

## Mesozoa

The Mesozoa are a little-known group of parasites of marine invertebrates, first reported in 1839, and later thought to be a link between the Protozoa and the Metazoa. Subsequent investigation has not substantiated this opinion, and the position of the Mesozoa in the animal kingdom still remains uncertain. Its members are thought by some zoologists to bear a strong resemblance to the sporocysts of trematodes, and for this reason they are regarded by some workers as degenerate forms of digenetic trematodes and conveniently placed in an appendix to the phylum Platyhelminthes. Other zoologists, however, consider the Mesozoa to represent a distinct phylum.

These animals are very small. The body lacks digestive, excretory, circulatory and nervous systems, and consists merely of a solid two-layered structure, but the two layers do not correspond with the ectoderm and endoderm of diploblastic animals. They have a complex life-history involving an alternation of asexual and sexual generations.

The known species of Mesozoa may be divided into two groups – the Dicyemida and the Orthonectida.

The Dicyemida are parasitic in the nephridia of cephalopods (squids and octopuses). They appear as a number of yellowish filaments, up to 7 mm long, attached to the wall of the nephridium, and are composed of not more than 25 cells, comprising an outer ciliated layer and a single, elongate, internal cell.

The Orthonectida are parasitic among the internal organs and tissues of marine turbellarians, nemertines, annelids and ophiuroids (brittle-stars), as well as gastropod and bivalve molluscs. The life history involves an asexual stage consisting of a multinucleate amoeboid plasmodium threading into the tissues of the host. At first the plasmodia reproduce by fragmentation, but later give rise asexually to males and females. The sexual forms are very small, elongate or oval animals, consisting of an outer ciliated epithelial layer and an inner cell-mass.

**Technique**
The study of mesozoans requires the microscopical examination of the histological and cytological details of their structure, and this can only be done satisfactorily with living specimens or with specimens fixed either in cover-glass smear preparations from the tissues of freshly killed hosts or in tissues from which sections are to be cut. Technical procedure for microscopical examination is beyond the scope of this work, and students are referred to articles on these parasites (Kozloff, 1965; Nouvel, 1947) as no text-books dealing with methods are yet available.

*References*
Kozloff, E. N. 1965. *Ciliocinita sabellariae* gen. and sp. n., an orthonectid mesozoan from the polychaete *Sabellaria cementarium* Moore. *J. Parasit.* **51**, 37–64.
Nouvel, H. 1947. Dicyémides (Pt. 1). *Arch. Biol., Paris*, **58**, 59–219.

# Nemertinea (Plate 4a)

The nemertines are free-living unsegmented worms in which the body is nearly always cylindrical and frequently extremely elongate. The entire surface of the body is covered with cilia and near the anterior extremity is a pore through which a tubular proboscis is extruded. The proboscis apparatus is used for capturing prey and in some groups it is armed with piercing stylets. The proboscis itself has no direct connection with the digestive tract, although in a small number of species the digestive tract opens by the proboscis pore (i.e. the pore *through* which the proboscis is extruded) and a separate mouth is lacking. Usually the mouth is situated ventrally some distance behind the proboscis pore and it leads into a digestive tract that runs the length of the body to the anus near the posterior extremity.

In most nemertines the sexes are separate and the eggs and sperms are produced in a series of little sacs. Each of these opens directly to the exterior through its own pore. The pattern of development varies and some marine species produce a free-swimming ciliated larva called a 'pilidium'.

The phylum is made up of about 550 species. The worms range in length from a few millimetres to several metres, although the majority are less than 200 mm long. *Lineus longissimus*, a species common in the North Sea, has been reported to attain a length of 30 metres.

Nemertines are mostly bottom-living marine animals, and they are found in sand, mud, gravel, under stones and amongst seaweeds in the intertidal zone and in shallow subtidal waters. A few pelagic species may be found in deep water. There are a few freshwater species, all of which are classified in the genus *Prostoma*. The genus *Geonemertes* encompasses a number of terrestrial species that are found in damp soils, under logs, etc., in tropical and subtropical regions. Although some nemertines have a restricted distribution, the group as a whole occurs all over the world.

## Collecting

Marine nemertines can be obtained by hand collecting in the intertidal zone and by dredging in shallow water. Similarly, freshwater species should be searched for amongst aquatic plants, under stones, etc., and they can often be obtained by sweeping water weeds with a fine-meshed hand-net.

## Preserving

Nemertines are difficult to preserve since, when subjected to the slightest irritation, they have a tendency to shed the proboscis or even disintegrate. Many workers recommend that the animals should be anaesthetized by adding crystals of chloral hydrate or magnesium sulphate to the water containing them (sea water for marine forms). Anaesthetization may take 6–12 hours. When insensitive, all except the hoplonemertines can be killed and fixed in 10 per cent formalin solution or in 30–50 per cent alcohol and stored in 3–5 per cent formalin or in 70–90 per cent alcohol. Formalin should not be used for fixing or preserving hoplonemertines because it dissolves the stylets that are borne on the proboscis in these animals.

If specimens are required for detailed histological investigation more elaborate fixation techniques are required. Satisfactory results have been obtained with Bouin's solution, but Mahoney (1973) recommends Heidenhain's 'Susa' (p. 129) and saturated mercuric chloride solution (p. 131). After fixing for 12 hours the animals should be transferred for a few hours to 70 per cent iodized alcohol (p. 131). Thereafter they can be stored in 70 per cent alcohol.

*Reference*
Mahoney, R. 1973. *Laboratory Techniques in Zoology*. London: Butterworths. 518 pp.


# Aschelminthes (Plate 5a–d, f, h–i)

The Aschelminthes are animals, mostly wormlike in appearance, that have an unsegmented or superficially segmented body and no distinct head. The body is generally covered by a distinct cuticle and the digestive tract is a straight or sometimes curved tube starting at an anterior mouth and terminating in a posterior anus. This group is extremely heterogeneous and indeed its component classes are often treated as separate phyla. One common feature of these animals is the presence of a particular type of body cavity called a pseudocoel, although this is also found in the Acanthocephala (p. 32), Entoprocta (p. 34) and in the Priapulida (p. 33) – a group included by some authorities in the Aschelminthes. Most Aschelminthes are small although some attain great lengths. The phylum includes free-living as well as many parasitic species and there are five classes: Rotifera, Gastrotricha, Kinorhyncha (also known as Echinodera), Nematomorpha and Nematoda.

## Rotifera (*Wheel-animalcules*) (Plate 5c–d)

The cylindrical or sub-cylindrical body of rotifers bears anteriorly a crown or corona of cilia, and in many species the beating of these cilia produces the illusion of a rotating wheel – hence wheel-animalcules. Rotifers are amongst the smallest multicellular animals and whilst a few may be about 2 mm in length the majority do not exceed 0·5 mm. In some species the cuticle is thickened to form a conspicuous encasement called a lorica, and some forms are enclosed in a gelatinous or cuticular tube with which mineral particles may be incorporated. The majority of rotifers are free-living but the class also includes a few epizooic and parasitic members. Rotifers are amongst the most common inhabitants of fresh water but some can also be found in damp terrestrial habitats (e.g. amongst mosses) and there are a few marine species. Most are solitary free-moving animals but there is also a number of sessile species and some of these form spherical swimming colonies.

**Collecting**

Pelagic forms can be taken by the usual methods of collecting plankton (p. 110). Species living attached to aquatic plants can be collected by placing the plants (especially those with finely divided leaves) with enough water to cover them in a glass jar illuminated on one side only. The jar should then be left undisturbed for two or three hours. The rotifers gradually come to the surface and congregate on the illuminated side, where they can be removed with a pipette. If the jar is left in direct sunlight the animals will not leave the plants. Moss-inhabiting species can be collected by examining the water squeezed out of moist moss samples under a binocular microscope or lens. Alternatively, samples can be teased apart in water under a low-power binocular, and materials such as mud, tree-hole debris etc., can be examined in much the same way. Many rotifers are able to withstand desiccation and they revive when wetted. Thus samples of moss, mud etc., for subsequent examination in the laboratory, can be allowed to dry out naturally – but they should not be heated – and stored in paper or plastic bags.

Plate 5 The pseudocoelomates, Aschelminthes, Priapulida, Acanthocephala a, a nematomorph or hair-worm; b, *Chaetonotus*, a gastrotrich; c, *Hydatina*, a rotifer; d, *Collotheca*, a rotifer with case; e, *Priapulus*, a priapulid; f, *Centropsis*, a kinorhynch; g, *Echinorhynchus*, an acanthocephalan, with proboscis detailed; h, *Ascaris*, a fusiform nematode; i, *Trichurus*, a part-filiform nematode.

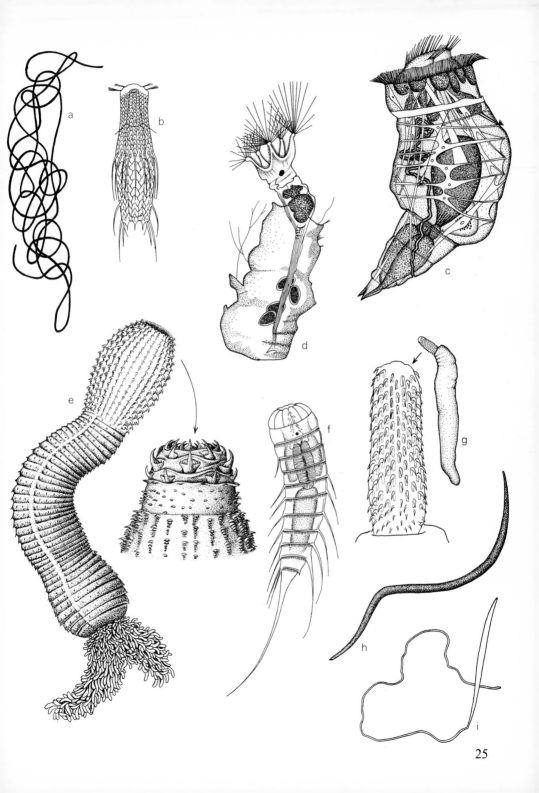

25

**Preserving**

For general taxonomic purposes the loricate rotifers can be killed and fixed by pipetting them directly into 10 per cent formalin solution. The illoricate species, however, have to be carefully anaesthetized before fixing, and Hanley (1949) recommends the use of benzamine hydrochloride. The time required for anaesthetizing with this substance varies from one species to another and is considerably affected by the pH of the water, acid water requiring more time and/or anaesthetic than alkaline water. The animals should not be left in the anaesthetic solution until ciliary movement has ceased, and a useful, but not infallible, indication of adequate anaesthetization is given when the animals start pushing themselves about without contracting. Assuming that the rotifers have been collected into about 8 ml of water in a shallow dish, for 'average' specimens, one or two drops of a 2 per cent aqueous solution of benzamine hydrochloride should be mixed with the water. After an interval of 20–30 minutes, another one or two drops of the anaesthetic solution are mixed with the water, and after a further interval of about 10 minutes the animals should be insensitive enough to be killed by the addition of a formalin solution (see below). It is seldom necessary to use more than six drops of the anaesthetic solution and anaesthetization seldom takes more than an hour.

Hanley also found empirically that rotifers could be anaesthetized very rapidly with a benzamine hydrochloride/cellosolve mixture (p. 125). With this solution the rotifers can be collected into very much less water, and for every 1 ml of water containing the animals from 0·125–0·5 ml of the anaesthetizing mixture should be added and stirred in quickly. Anaesthetization is extremely rapid and some species become sufficiently insensitive for killing within three minutes.

After anaesthetization the rotifers can be killed and fixed by adding and stirring in a small quantity of 10 per cent formalin. After killing, specimens should be washed about six times in fresh 5 per cent formalin in order to remove all the benzamine hydrochloride. Material can then be stored in 2–5 per cent formalin.

*Reference*
Hanley, J. 1949. The narcotization and mounting of Rotifera. *Microscope*, **7**, 154–159.

# Gastrotricha (Plate 5b)

The gastrotrichs are minute free-living Aschelminthes. The more or less elongated body has a flattened ventral surface and a convex dorsal surface, and in many species the head is delimited by a slightly constricted neck region. Cilia occur only on the head and on the ventral part of the trunk. There are two orders, one (the Macrodasyoidea) being made up entirely of marine species and the other (Chaetonotoidea) including both marine and freshwater forms. The marine species occur chiefly in sand in coastal waters but they have also been found amongst seaweeds. The freshwater forms have for the most part benthic habits and live amongst vegetation in ponds and lakes and in moss pools and bog water. Comparatively little is known of their geographical distribution. Most marine species have been taken from European coasts but many freshwater genera appear to be cosmopolitan.

### Collecting

Gastrotricha are generally collected by simple washing and sieving techniques. Vegetation can be rinsed with fairly large quantities of water and the washings strained through a fine sieve or through bolting-silk. Bottom-dwelling species can be collected by carefully scooping up the superficial layers of mud and debris and allowing the material to stand for several hours in a jar of water. Material at the substratum/water interface can then be sucked up with a long pipette and examined in a shallow dish under a binocular microscope.

Boaden (1963) collected marine interstitial species by nearly filling Brevit jars of 2·5 litre capacity with samples of beach sand. The jars were then topped up with sea water to a level about 10 mm above the surface of the sand and left uncapped and undisturbed in the laboratory for three days. Surface sand in the jars was then transferred to a beaker and swirled up vigorously in sea water or in a 6 per cent solution of magnesium chloride. As soon as the sand had nearly settled the solution was strained through fine plankton netting and the entrapped animals washed off into a shallow dish. For other extraction methods see p. 98, meiofauna.

### Preserving

For gross morphological and taxonomic work fixation in 10 per cent formalin is generally adequate. For freshwater forms Pennak (1953) recommends 2 per cent osmic acid as a fixative. The living specimens are placed in a drop of water on a microscope slide and the slide inverted for 5–10 seconds over the bottle of osmic acid. Some species (especially the marine forms) contract strongly when fixed and require prior anaesthetization. Various anaesthetics (including chloral hydrate and even cocaine) have been used with varying success and trials with newer materials such as MS 222 (p. 124) and propylene phenoxetol (p. 124) could be worthwhile. Gastrotricha can be stored in 70 per cent alcohol or in 5 per cent formalin.

*References*
Boaden, P. J. S. 1963. Marine Gastrotricha from the interstitial fauna of some North Wales beaches. *Proc. zool. Soc., Lond.* **140**, 485–502.
Pennak, R. W. 1953. *Fresh-water Invertebrates of the United States.* New York: Ronald Press Company. 769 pp.

## Kinorhyncha or Echinodera (Plate 5f)

The Kinorhyncha are small Aschelminthes devoid of cilia and with a regularly but superficially segmented body made up of 13–14 joints. The head region is retractile and both this region and the trunk are covered with spines. The Kinorhyncha are exclusively marine animals and are all less than 1 mm in length. They have benthic habits and are generally found in muddy bottoms or amongst algae. They do not swim but move about in a worm-like fashion with the aid of posteriorly directed spines (scalids) on the head region. Most of the known species have been taken off European coasts but there are also isolated records from other widely separated localities and they probably have a world-wide distribution.

### Collecting
Kinorhyncha have generally been collected by simple washing, sieving and flotation procedures (p. 98). Those living on algae can be taken by washing the plant material in fairly large quantities of sea water and filtering the washings through a fine-mesh sieve or bolting-silk. Similarly those living amongst mud and slime can be collected by washing the upper layers of the substratum through a fine sieve. When large quantities of organic and/or inorganic debris are collected on the sieve a further separation can be made by using flotation procedures (p. 98).

### Preserving
For general taxonomic purposes Kinorhyncha can be fixed and stored in formalin solutions – 10 per cent formalin for fixation and 2–5 per cent formalin for storage. It is an advantage to kill the animals with their head region fully extended. This can usually be done by fixing the animals on a microscope slide while exerting slight pressure on the cover-glass, or by placing them for a short while in distilled water before transferring them to the fixative.

# Nematoda (Plate 5h, i)

The nematodes are worm-like Aschelminthes and in most species the elongated cylindrical body is tapered at both ends. The majority of these animals are free-living but the group also includes many animal and plant-parasitic forms. Most of the free-living nematodes are small species less than a millimetre in length. However, some free-living marine species can attain lengths of 50 mm and some animal parasites, such as species of *Ascaris*, can be more than 300 mm long.

This class contains some of the most widespread and numerous of all multicellular animals. They have a world-wide distribution and have exploited virtually all terrestrial and aquatic habitats. They live in fresh and salt water and the habitats of the free-living terrestrial forms extend from the lowest intertidal zones to high mountain elevations. Organic debris of all sorts can support vast populations and they are the most numerous multicellular animals in nearly all types of soil. The parasitic forms display all sorts of associations and they attack virtually all groups of plants and animals. Many are important as pests of crops and as parasites of domestic animals, and a considerable number are responsible for human diseases.

## Collecting

A few of the free-living species that are found in soil, fresh water, or in sand and amongst seaweeds on the shore are large enough to be seen easily and can be picked up with forceps or with a moistened paint brush. However, the majority of the free-living species are obtained by special methods. The inhabitants of mud and sand can be obtained by washing and sieving techniques (p. 98). These methods can also be used for collecting soil-dwelling species but materials containing an appreciable amount of organic matter are best dealt with by means of the Baermann funnel technique (p. 95).

Plant-parasitic species can often be collected by teasing apart pieces of infested material in water under a binocular microscope or lens, but the Baermann apparatus can also be used, small pieces of infested material being placed in the muslin bag. A number of rather elaborate procedures (designed mainly for quantitative work) has been devised for extracting nematodes from soil and plant material. Many of these techniques are described by Goodey (1963), but most of the procedures require good laboratory facilities and are rather beyond the scope of the general collector.

Nematodes living as internal parasites can be collected by the procedures outlined on pp. 119–121.

**Preserving**

The larger free-living nematodes can be killed and fixed by immersing them in hot or cold 3–5 per cent formalin solution or in hot 70–90 per cent alcohol. Small numbers of small free-living species can be placed in a drop of water on a microscope slide (or any suitable slip of glass) and killed by heating the glass over a flame. The specimens can then be transferred to 3–5 per cent formalin.

When large numbers of small free-living species have to be dealt with, they can be collected into a small tube or concentrated in a centrifuge tube with a small amount of water. The animals can then be killed by immersing the tube containing them for at least two minutes in a vessel of water heated to 60–65°C. An amount of 20 per cent formalin or 'double strength' T.A.F. (p. 132) equal in volume to the water containing the nematodes should then be added. When dealing with marine species the formalin and T.A.F. should be made up in sea water. Nematodes fixed in this way can be left indefinitely in either of the fixative solutions.

When facilities are poor, marine nematodes can be fixed in bulk by placing samples of seaweeds (including the holdfasts and surrounding debris), mud, sand and so on in jars with a small amount of sea water. An amount of commercial formalin sufficient to produce an approximate 10 per cent solution of formalin in sea water can then be added to the samples.

Owing to the resistant nature of their cuticle hot fluids are almost indispensable for killing and fixing nematodes parasitic in animals. The worms should be washed by shaking them up in a 1 per cent salt solution, and killed and fixed by dropping them separately into hot 70–90 per cent alcohol (important, see p. 18) or hot 3–5 per cent formalin. The fixative fluids should be kept steaming, but not quite boiling, and a temperature of 70–80°C is usually high enough. Hot water is not recommended for killing parasitic forms as it is liable to cause excessive stretching of the cuticle. After the fixative fluid has been allowed to cool the nematodes should be transferred for storage to fresh fluid of the same kind.

When specimens are small and numerous it may be easier, after draining off as much as possible of the washing fluid, to pour the hot fixative over them. When cool the whole may then be poured into a tube or bottle and allowed to settle. The used fluid can then be poured off and replaced by fresh solution.

*Reference*

Goodey, J. B. 1963. *Laboratory Methods for Work with Plant and Soil Nematodes*. Ministry of Agriculture, Fisheries and Food Technical Bulletin No. 2 (4th edn), London: H.M.S.O. 544 pp.

# Nematomorpha (*Hair worms*) (Plate 5a)

The Nematomorpha are Aschelminthes which as larvae are parasitic in certain arthropods. However, the adults, which are extremely long (often attaining a length of 1 m) are free-living, and predominantly aquatic, although some species are found in damp soil. The group has two major divisions: the Gordioidea, which includes all the freshwater and terrestrial species, and the Nectonematoidea which comprises a small number of marine pelagic species all classified in the genus *Nectonema*.

In general structure the Nematomorpha resemble the nematodes, but, both in the adults and juveniles, the digestive tract is more or less vestigial. The adults probably do not feed at all and the juveniles apparently absorb nutritive material from the host directly through their body wall. The sexes are separate and the eggs are deposited in water. On hatching, the larvae seek an arthropod host either in or on the edge of the water. Common hosts are beetles and various Orthoptera (crickets, grass-hoppers, etc.) although centipedes and millipedes may also be infested. The larvae of *Nectonema* species are found in hermit crabs and in true crabs. After penetration, the larvae enter the haemocoel of the host and remain there until they are almost fully developed adults. Species parasitizing terrestrial arthropods emerge only when their hosts are near water. Sexual maturity is attained during a short free-living phase. The group has a world-wide distribution.

**Collecting**

The adult worms are often found tangled together in great numbers in very shallow water and, less frequently, in damp soil, and they are easily picked up by hand. In deeper water they can be taken with a hand-net. Juveniles can be collected by dissecting suitable arthropod hosts under a binocular microscope.

**Preserving**

For general taxonomic purposes the worms can be fixed in cold 3–5 per cent formalin solution and stored in the same fluid. For the study of fine anatomical detail corrosive acetic (p. 131) is a better fixative Material should be fixed for about 30 minutes. Thereafter the fixative should be removed by washing with several changes of 70 per cent alcohol, or by treatment with iodized alcohol (p. 131). Material fixed in this way should be stored in 70–90 per cent alcohol.

# Acanthocephala (*Thorny headed worms*) (Plate 5g)

The acanthocephalans are endoparasitic worms which in their pseudo-coelomate structural plan have many features in common with the Aschelminthes, particularly the Nematoda. However, they also have many 'non-aschelminthine' as well as a number of unique characters and they are set aside by most zoologists as a separate phylum. The adult body is cylindroid although its precise form varies, some species being short and stumpy, some fusiform and others long and slender. The body is divisible into two regions – a short anterior presoma (made up of the proboscis and neck) and a longer and stouter posterior trunk that may be superficially segmented. The proboscis is retractile and armed with stout recurved hooks. There is no trace of a digestive tract. As adults the worms range in length from about 1·5 to 500 mm, although most species are less than 25 mm long.

Adult acanthocephalans are parasites in the digestive tracts of vertebrates, mainly fish, birds and mammals. In general, the smaller species are found in fish and the larger in avian and mammalian hosts. Immature stages of the worms are found in intermediate arthropod hosts (mainly Crustacea and insects) which fall prey to the hosts of the adults. There are about 500 species and the group is world-wide in distribution.

## Collecting
Procedures for collecting intestinal parasites are outlined on p. 119.

## Preserving
Freshly collected material should be cleaned with a moistened paint brush, for if left soaking in water or salt solution the worms tend to swell. Material can be fixed in hot 3–5 per cent formalin or hot 70–90 per cent alcohol (important, see p. 18), using the procedures recommended for endoparasitic nematodes (p. 30). Mahoney (1973) recommends fixation in warm (about 50°C) corrosive acetic (p. 131). The animals should be left in this solution for about 30 minutes and then washed several times in iodized alcohol (p. 131) before being stored in 70 per cent alcohol. If possible Acanthocephala should be killed with the proboscis extruded. This can generally be done by gently pressing the fresh material between two slips of glass until the proboscis is fully extended and then, without releasing the pressure, allowing the fixative fluid to run in between the slips. The pressure can be released after the material has been in contact with the fixative for 2–3 minutes.

32

*Reference*
Mahoney, R. 1973. *Laboratory. Techniques in Zoology*. London: Butterworths. 518 pp.

# Priapulida (Plate 5e)

This small phylum comprises a group of marine worms in which the body is divisible into a shorter anterior proboscis region and a posterior trunk. The proboscis is somewhat barrel-shaped and ornamented with longitudinal rows of papillae, and the terminal mouth, which invaginates into this region, is encircled by a formidable armature of spines. The trunk region is superficially segmented and covered with small spines and tubercles. The phylum comprises only four species which are classified in three genera: *Priapulus*, *Halicryptus*, and *Tubiluchus*. The genus *Priapulus* is characterized by the presence of one (*P. caudatus*) or two (*P. bicaudatus*) caudal appendages. These appendages, which arise from the posterior end of the trunk, are hollow stems beset with hollow vesicles. Their function is not known. *Priapulus* species are relatively large worms, often exceeding 80 mm. in length, which live buried in sand or mud in the littoral zone of colder seas. The only known species of the genus *Halicryptus* – *H. spinulosus* – is rather smaller, ranging in length from about 17–30 mm. Its habitats are similar to those of *Priapulus* species and it is widely distributed in the colder seas of the northern hemisphere.

### Collecting
*Priapulus* and *Halicryptus* can be obtained by digging within the intertidal zone or by dredging offshore in water of moderate depths. Methods of collecting interstitial species are outlined on p. 98.

### Preserving
Priapulids should be anaesthetized before fixing. A 7 per cent solution of magnesium chloride is generally effective, although with large specimens anaesthetization may take some time. Afterwards the animals can be fixed and stored in 5 per cent formalin solution. Alternatively, anaesthetized material can be fixed for about 24 hours in Bouin's fluid (p. 128) and then placed in 50 per cent alcohol for 1 hour before being transferred for storage to 70 per cent alcohol.

## Entoprocta (Plate 6a)

This small phylum comprises a group of colonial or solitary aquatic animals, which are marine with the exception of a very few freshwater species. The Entoprocta were previously included with the Bryozoa (p. 34), from which they differ in important respects.

The individuals are small (length 0·25–3·00 mm). The gut is U-shaped, but the anus always opens within the tentacular crown, which is not retractable within an introvert as in the Bryozoa, and there is no coelom.

Colonies and populations are formed by asexual budding, and are also produced sexually from the metamorphosis of ciliated larvae. Entoprocta are found encrusting various substrata, and the solitary forms are found as ectoparasites of marine worms and sponges.

### Collecting

Entoprocta are usually inconspicuous, and are found when large masses of seaweed, or of other animal phyla, are being examined. Under a low-power microscope, seaweeds, hydroids, Bryozoa, etc., may be seen to be associated with colonial forms, which are recognised by the rapid movements of an erect 'stalk' portion with its rounded terminal 'calyx' which comprises the viscera and tentacles. Ectoparasitic forms are found on the gills and limbs of marine worms, or are associated with sponges, molluscs and echinoderms.

### Preserving

Anaesthetization is absolutely essential, and the same methods may be used as for Bryozoa (p. 39). Specimens should then be fixed in Bouin's fluid (p. 128), washed and preserved in 70–90 per cent alcohol.

## Bryozoa (Polyzoa or Ectoprocta) (Plate 6b–h)

With very few exceptions the members of this exclusively aquatic group are sedentary, colonial animals. The great majority are marine, but there is a distinct group of freshwater species. Whole colonies may be large (over 1000 mm in diameter), but generally they range from 10–100 mm in size. There are about 4000 known living species. The units (zooids) composing a colony are small (average length 0·5 mm); larger colonies therefore include many thousands of zooids. Superficially, zooids of some species may resemble hydroid polyps, but the Bryozoa are considerably more complex in structure than coelenterates. The body wall is formed from an epidermis that secretes an overlying cuticle. In most groups the epidermis also lays down calcification beneath the cuticle. The body cavity is a coelom in which is suspended the 'U'-shaped gut. Complex extensions and expansions of the coelom may also occur on the outer side of some of the calcified body walls; these are connected to the visceral coelom through special pores. Similar pores also supply communication among the coeloms of adjacent zooids, so that there is transfer of nutrients for growth, repair of damage, etc., in any direction throughout the life of the entire colony. The mouth is situated near the centre of a retractable circular or horse-shoe shaped structure – the lophophore – which bears numerous ciliated tentacles. The lophophore with its tentacles can be retracted within an introvert of body wall, and is protruded for feeding through an orifice in the body wall by means of complex hydrostatic mechanisms requiring several muscle systems. Part of the gut is also ciliated, and the anus is located near the mouth but always outside the tentacle crown (cf. Entoprocta, p. 34). Each zooid possesses a small central nervous system with a ganglion lying withing the coelom below the lophophore. A dermal nerve network and nervous connection among zooids is known in some species. There are no special respiratory organs, and no circulatory or excretory systems.

Most Bryozoa are polymorphic; the greater number of zooids in a colony may have a feeding function, but others may be partially or wholly modified for additional or entirely different functions, such as attachment, support, defence, cleaning, etc.

Colonies of both marine and freshwater species are formed by asexual budding of series of zooids. In a few cases new colonies may be formed by fragmentation, but usually they are produced by sexual reproduction. Colonies may also be produced, in freshwater and estuarine forms, by special resistant zooids, or, in one freshwater group, by internally budded, resistant statoblasts. These may be seen as small, dark brown, round or oval bodies within the semi-transparent colonies. Ova and spermatozoa are produced in ovaries and testes formed seasonally from peritoneal cells. Most bryozoan zooids are hermaphrodite, but some species produce specialized, male zooids. Sperm are released through pores at the tips of one pair of tentacles; fertilization is apparently internal. In some marine Bryozoa the fertilized ova may be released through a coelomic pore at the base of this pair of tentacles into a ciliated 'tentacular organ', and then expelled into the sea where they develop into larvae that have a 'bivalve' shell and a ciliated gut. These swim, feed and grow in the plankton before settling on a suitable substratum and undergoing metamorphosis. In another type of development the ova are released into a special internal or external brood chamber for development. External brood chambers are produced in several different ways and usually have complex calcified walls. Larvae released from the brood chambers have no gut, but a large 'yolk' and cilia. They usually have a short free-swimming life (8 hours to 5 days), before settlement and metamorphosis into the first zooid of a new colony. In freshwater Bryozoa the ova develop in an embryo sac and a small ciliated colony of 2–3 zooids is released. This has a short free-swimming life.

The phylum is usually divided into three classes: the Stenolaemata, Gymnolaemata and Phylactolaemata. The Stenolaemata are marine, the Gymnolaemata almost exclusively marine, and the Phylactolaemata exclusively freshwater. The Stenolaemata and one large group of the Gymnolaemata have calcified body walls and a long fossil record. A smaller group of the Gymnolaemata and the Phylactolaemata do not have calcified walls, but fossils of the Gymnolaemata are well known, particularly those of a group that live by boring within the calcified structure of shells, etc.

Bryozoa are cosmopolitan in distribution. Freshwater forms are found in high mountain lakes, and in rivers throughout the world. Marine forms are amongst the commonest sedentary animals in shallow coastal and deeper shelf waters of all seas, and very specialized species occur at abyssal depths. Colonies are usually attached to a fairly firm substratum, but some are capable of colonizing unstable, sandy habitats, and a few are part of the interstitial sand fauna. Most colonies form lacy or massive encrustations, plant or hydroid-like tufts or rigid coral-like or foliaceous masses. Some forms, both erect and encrusting, are fleshy and gelatinous. A few species are consistently associated with Crustacea, some living on the limbs and antennae, others on the mollusc shells inhabited by hermit crabs. Large specimens of crabs from deeper water often have Bryozoa encrusting the carapace.

Freshwater Bryozoa are gelatinous in appearance and form delicate ramose encrustations, or large fleshy masses on stones, submerged tree roots, logs and the stems and leaves of water plants.

Plate 6 Entoprocta (a) and Bryozoa (b–h)
a, *Pedicellina*, an entoproct, zooids 0·5 mm; b, *Plumatella*, colony, zooids 2 mm; c, *Crisia*, erect, jointed colony, zooids 0·5 mm; d, *Flustra*, erect, bilaminar, flexible colony, zooids 0·5 mm; e, *Sertella*, erect, unilaminar, rigid colony, zooids 0·5 mm; f, *Pentapora*, bilaminar, anastomosing, rigid colony, zooids 0·5 mm; g, *Bugula*, flexible, unilaminar colony, zooids 0·5 mm; h, *Flustrellidra*, encrusting, flexible colony on *Fucus*, zooids 1 mm.

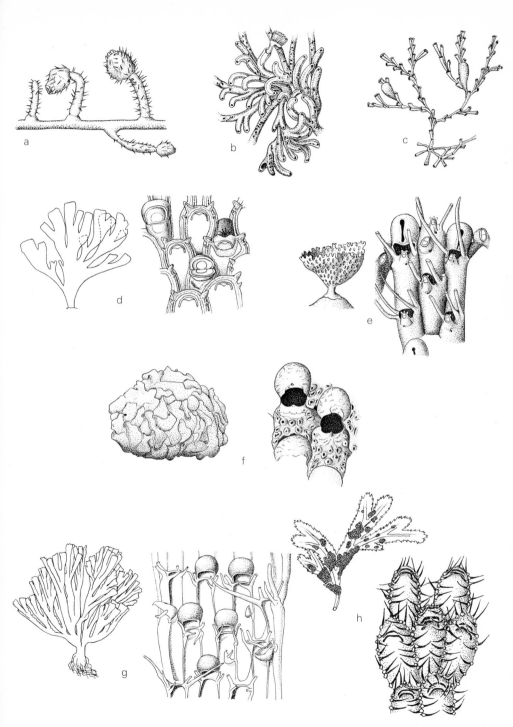

Bryozoa are important fouling organisms. Marine species are often completely resistant to anti-fouling paints and form an insulating layer upon which the larvae of other fouling animals such as barnacles can settle and grow. Large colonies are found on harbour installations and on the bottoms of boats. Both marine and freshwater species frequently clog the water-intake screens of electricity power-stations.

## Collecting

Freshwater specimens can be found by examining suitable substrata with a hand-lens. Portions of large colonies growing on rock or large stones should be removed with a knife; smaller specimens are best collected intact with their substratum.

Marine Bryozoa are often mistaken for hydroids, corals, colonial seasquirts or seaweeds. They may be distinguished, using a hand-lens, by the complexity of structure and regularity of arrangement of units. Each zooid orifice (about 0·1 mm in diameter) forms part of a definite, colony-wide pattern, which is emphasized by the frequent occurrence of spines and polymorphs. Colonies from cold or temperate waters are usually greyish-white to orange-brown in colour, those from warm and tropical waters are often brilliantly coloured, ranging from purple and scarlet to yellow and green. The lophophores are often coloured, as are most embryos. External brood chambers are often prominent and globular, and when embryos are present, form coloured zones in colonies which are easily seen with a hand-lens.

Marine species in shallow water may be found by examining rocks, shells and seaweeds. Worn shell is one of the most frequent substrata and colonies of as many as 30 species may be found encrusting a single large shell. Colonies often completely cover the lower parts of the fronds of *Fucus* and *Laminaria*, and small erect species are often found on the holdfasts of these seaweeds. Fairly large specimens (about 100 mm in size), often attached to a shell or small stone, are usually washed up after storms and may form a major part of the strand debris in some seasons. In deeper water, the best method of collecting is by free diving. Many erect and encrusting species cover the walls and roofs of underwater caves, and grow from beneath overhanging rock surfaces. Most man-made substrata such as pottery, glass, metal and plastic objects are also covered by colonies. Wherever possible, whole colonies should be detached together with their immediate substratum. Each large colony, or group of small colonies should be kept separately in a polythene bag of sea water, so that the delicate terminal growing edges are not damaged. Material collected by dredging is often broken, but should be treated in the same manner. Dredged samples from muddy and sandy bottoms should be examined under a low-power microscope for small, specialized colonies (about 1·0–5·0 mm in diameter).

Wherever possible, observations on colour and associated fauna should be made. Colonies should be placed in dishes of aerated water and left undisturbed until the tentacle crowns are expanded. Small predators, such as pycnogonids and nudibranch Mollusca should be removed and preserved. Using a low-power microscope, colony behaviour, such as patterns of water movement and functions of polymorphs, may be observed.

Some free-swimming larvae may be taken with other plankton in a fine mesh tow-net. Others may be released from the brood chambers of colonies kept in aerated water and exposed to light. If suitable substrata are provided the larvae may settle and metamorphose.

### Preserving
After observing the living colonies, an attempt should be made to anaesthetize them using 1 per cent stovaine (p. 126), eucaine (p. 125) or similar agent, which should be added to the dishes of water drop by drop, with intervals of 10–15 minutes between the first two doses, and 5 minutes between subsequent doses. Slow addition of menthol crystals, or of epsom salts is an alternative method. When the tentacle crowns cease to respond to touch, the colonies may be killed by the rapid addition of commercial formalin or Bouin's fluid (p. 128). Marine species with calcified walls must always be washed gently in fresh water and stored in 70–90 per cent alcohol. Other forms, including the freshwater species, may be stored in 5 per cent formalin solution. If preservatives are not available, or the specimens are very large, species with calcified walls may be washed in several changes of fresh water and then dried.

Larvae should be transferred to a dish with a pipette, anaesthetized, and killed with corrosive sublimate (p. 131) or Bouin's fluid (p. 128). The solution should then be pipetted off, the larvae washed with water and preserved in 70–90 per cent alcohol.

## Phoronidae

The phoronids are sedentary marine worms that inhabit chitinous tubes of their own secretion. Their cylindrical bodies range in length from about 6–200 mm, most species being less than 100 mm long, and as in the case of the Bryozoa (p. 34) and Brachiopoda (p. 40), to which they have close affinities, the mouth is surrounded by a tentacular feeding

organ, or lophophore. Phoronids occur in the littoral zone and in shallow subtidal waters in tropical and temperate regions. Their tubes are found buried in sand or mud, or attached, either singly or as tangled aggregations, to rocks, pilings, shells and other objects. Some species burrow into calcareous shells, but within these burrows they are surrounded by their secreted tube. Eggs develop into free-swimming larvae called actinotrochs. After several weeks of a planktonic existence the larvae sink to the bottom and undergo a very rapid metamorphosis, following which the young worms immediately begin to secrete their tubes. At least one species, the widely distributed *Phoronis ovalis*, which occurs in aggregations, reproduces asexually both by budding and by a process of transverse fission. The phylum is made up of about 15 species classified in two genera: *Phoronis* and *Phoronopsis.*

**Collecting and preserving**
The methods recommended for tubiculous marine polychaetes (p. 47) are appropriate.

# Brachiopoda

These exclusively marine animals resemble bivalved molluscs in possessing a calcareous shell made up of two valves. However, the resemblance is quite superficial. In the brachiopods the two valves are dorsal and ventral in position with reference to the enclosed body – in the bivalved molluscs they are lateral, although in most species the ventral valve lies uppermost in the normal orientation of the attached animal. The ventral valve of brachiopods is typically larger than the dorsal and in some species the apex is drawn out to form a curved perforated beak. The body proper of the animal occupies only the posterior part of the space between the valves, but extensions of the body wall known as the mantle lobes line those parts of the valves not occupied by the body. The space between the mantle lobes is occupied by a large tentacular feeding organ, the lophophore, which is thought to correspond to the lophophore of bryozoans (p. 34) and phoronids (p. 39). The tentacular arms of the lophophore have ciliated grooves and the beating of the cilia sweeps minute organisms into the mouth that is situated between the bases of the arms.

The phylum is divided into two classes: Inarticulata and Articulata. In the Inarticulata the valves are held together by muscles only and they can be opened very widely, while in the Articulata the valves are locked together posteriorly by a tooth and socket arrangement that greatly restricts their gape.

Most brachiopods are attached to the substratum by a stalk-like extension of the ventral body wall called the peduncle. Amongst the Inarticulata, the peduncle of *Lingula* and *Glottidia* species (family Lingulidae) is long and muscular. These brachiopods live in vertical burrows in sand or mud in tropical and subtropical waters, and when feeding the extreme anterior part of their shells protrude with the valves slightly agape through the slit-like openings of the burrows. In most other brachiopods the peduncle is short, thick and devoid of muscle fibres. It issues from the dorsal side of the ventral valve, either through a hole in its upturned apex or through a notch in the hinge-line, so that when attached to the substratum the animals are held in a more or less horizontal position with their larger ventral valves uppermost. In some genera of both classes the peduncle is completely absent, and in these animals the ventral valve is cemented directly to the substratum.

With few exceptions the sexes in brachiopods are separate and the eggs or sperms, which are discharged into the body cavity, reach the exterior by way of excretory tubules. Fertilization appears to occur at the time of spawning and the embryo develops into a free-swimming larva. The larvae of the Inarticulata resemble minute brachiopods, and as the shell is secreted by the mantle folds the more mature individuals gradually sink to the bottom. In the Articulata the larvae do not resemble the adults so closely. They have a relatively large anterior ciliated lobe and undergo a metamorphosis after a short free-swimming life.

Brachiopods are world-wide in distribution although they occur in great abundance in only a few areas; for example, off southern Australia, New Zealand and parts of Japan. They occur in less variety in the northern Atlantic, Arctic and Antarctic oceans and in the Mediterranean, and their general pattern of distribution suggests a preference for cooler rather than truly tropical seas. Although a few species are known from abyssal depths, the majority live on continental shelves at depths of 200–300 m. They are rarely found above low tide level. There are only about 300 living species known but in the past the group was extremely abundant, and about 30 000 fossil species have been described. They abounded in Palaeozoic seas, but during the Mesozoic era, although still abundant, they began to lose ground to the molluscs. *Lingula* and *Crania* share the distinction of being the oldest known genera of present-day animals, both dating back to the Ordovician period.

**Collecting**
Brachiopods are usually obtained by dredging or trawling (see p. 105).

**Preserving**

The animals should be anaesthetized by gradually adding small quantities of alcohol to the sea water, but the amount added should not exceed 10 per cent of the volume of the sea water containing the animals. Anaesthetics such as MS 222 (p. 124) and propylene phenoxetol (p. 124) may also prove to be effective. After anaesthetization the animals can be fixed and preserved in 70–90 per cent alcohol, but a small chip of wood should first be placed between the valves of the shell to allow the alcohol to penetrate to the soft parts. If the material is required for histological preparations the animals should first be anaesthetized, and then killed by gradually replacing the sea water with 5 per cent formal saline solution since alcohol tends to destroy the fine structure of the cells.

## Sipuncula (*Peanut worms*) (Plate 7h)

The sipunculans are unsegmented marine worms in which the body is divided into two regions – a slender anterior region called the introvert and a larger, thicker posterior trunk. The introvert, which can be contracted into the anterior portion of the trunk, has the mouth at its anterior extremity which is generally encircled by tentacular outgrowths. Sipunculans have a well-developed recurved digestive tract suspended in a large body cavity (coelom) with the anus located mid-dorsally on the anterior part of the trunk or on the posterior part of the introvert. The animals vary considerably in size, the smaller species being about 3 mm and the larger about 600 mm in length. The majority are between 150–300 mm long.

Sipunculans are exclusively marine and are found in all seas from the intertidal zone to considerable depths. They are all sedentary animals, and may be found in burrows in sand, mud and gravel, in crevices in rocks and corals, amongst holdfasts of seaweeds, and indeed in almost any protected situation. Their feeding habits are not fully understood. They may be ciliary feeders, but some species at least ingest considerable quantities of sand and mud from which organic material could be digested.

**Collecting**

The methods employed for collecting these animals are similar to those described for marine bristle worms (p. 47).

42

**Preserving**

The animals should be anaesthetized before being killed to ensure that they are preserved with their introvert everted. This can be done by adding small quantities of alcohol, propylene phenoxetol, or magnesium chloride crystals to the sea water containing them. When the animals no longer respond to probing they should be straightened out in a flat dish and fixed in 70–90 per cent alcohol or 4 per cent formalin solution. Slight pressure should be exerted on the body with a glass slip (until they become rigid), to keep the introvert extended. The animals should be fixed for at least 12 hours and then transferred to fresh solution for storage.

# Echiura (*Spoonworms*) (Plate 7f, g)

The Echiura are unsegmented, bilaterally symmetrical, coelomate, marine worms in which the body is divided into a muscular sausage-shaped trunk and an extendable anterior proboscis. The body surface may be smooth, or ornamented with annulae or papillae and usually has a pair of ventral setae, and sometimes one or two rings of anal setae. The animals range in size from a trunk length of a few millimetres to as much as 600 mm, and the proboscis may be only a fraction of the trunk length or may be many times the length. Unlike the sipunculan introvert, the echiuran proboscis cannot be retracted within the body. The shape of the proboscis is quite variable – it may be broad and flattened with the edges rolled under to form a channel, or long and slender, often with the distal region expanded. In other species it is bifurcate to a greater or lesser extent. The mouth is situated at the base of the proboscis, and a long coiled gut leads to the anus that opens at the extreme posterior end of the body.

The name 'spoonworm' is derived from the feeding activity exhibited by some species in which the scoop-like proboscis is used to pass sediment and detritus to the mouth. The proboscis is extended onto the sediment and the food particles are carried towards the mouth by the action of surface cilia. In the genus *Urechis* food is obtained in a rather different manner – a mucus net is secreted across the burrow by the animal and used to trap bacteria and suspended particles. From time to time the mucus trap is ingested by the spoonworm.

At one time the echiurans were considered to form a link between annelids and holothurians and were placed, together with priapulids and sipunculans, in the class Gephyrea, within the phylum Annelida. More recent work on the embryology has shown them to be quite different from annelids and has led to the establishment of a separate phylum. The group is divided into three orders: Heteromyota, Echiuroinea and Xenopneusta. At the present time there are about 130 known species. A detailed account of the group is given by Stephen & Edmunds (1972).

Spoonworms are exclusively marine, with the exception of a few brackish-water species, living mostly in burrows or beneath stones in soft sediments. Other species inhabit cracks and crevices in rocks or coral growths. They have a world-wide distribution occurring in tropical, temperate and polar seas, living mainly in shallow littoral waters but also at abyssal and intermediate depths.

## Collecting

Echiurans living in soft sand and mud can be collected by careful digging and sifting. Burrowing species can often be located by the presence of holes in the sediment, while in others the proboscis may be seen spread on the surface. The animals are very delicate and must not be removed by pulling the proboscis since this will break off leaving the trunk region behind. If the proboscis is detached during collection it must be kept and preserved with the trunk as it is an important taxonomic structure. Spoonworms living in rocks and corals can sometimes be made to vacate their crevice by the injection of a weak solution of formalin.

## Preserving

To avoid distortion of the body due to the contraction of the musculature, the animals should be anaesthetized before fixation. This can be achieved by the addition of a few crystals of menthol, or the dropwise addition of 90 per cent alcohol to the water containing the animals, or by immersion in a 7 per cent solution of magnesium chloride, or 1 per cent solution of propylene phenoxetol. When the animals no longer respond to probing they can be killed and fixed in 5 per cent neutral formalin solution and then transferred to 70 per cent alcohol for preservation.

Plate 7 Annelida, Sipuncula, Echiura

a, *Aporrectodea*, an oligochaete, earthworm; b, *Hydroides*, calcareous tubes of serpulid polychaete on a shell; c, *Neanthes*, an errant nereid polychaete; d, *Neoamphitrite*, a sedentary terebellid polychaete; e, *Hirudo*, a leech; f, *Echiurus*, an echiuran or spoonworm; g, *Bonellia*, a spoonworm with bifid proboscis; h, *Dendrostomum*, a burrowing sipunculan or peanut worm.

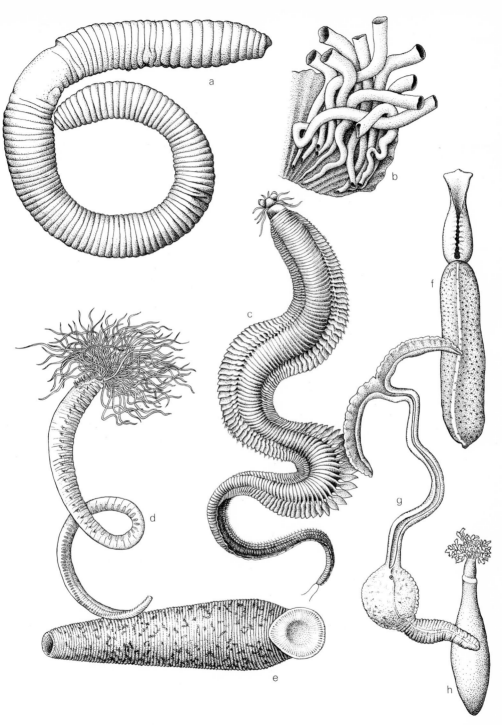

45

*Reference*
Stephen, A. C. & Edmonds, S. J. 1972 *The Phyla Sipuncula and Echiura.* London: British Museum (Natural History). 528 pp.

## Annelida (*Ringed worms*) (Plate 7a–e)

The animals in this phylum (bristleworms, earthworms and leeches) have a body made up of a linear series of similar segments. This condition, referred to as metameric segmentation, is not merely superficial but effects many internal structures, and even the body cavity (coelom) of these animals may be divided by transverse septa which mark the boundaries of the external segments. The phylum is made up of three classes: Polychaeta*, Oligochaeta and Hirudinea.

## Polychaeta (*Bristleworms*)

The majority of polychaetes have on each segment a pair of lateral bristle-bearing lobes called parapodia, and they usually have a distinct head which bears a number of appendages. Although there are a few freshwater species, the bristleworms are found mainly in marine or estuarine habitats. The class is made up of a large number of families which can be allocated to one of two ecological or life-form groups – Errantia or Sedentaria. Although the difference between the two extremes of the life-form spectrum is very striking, these two groups are not well defined, because there are many species and stages which have an 'intermediate' mode of life.

The errant polychaetes include the free-moving and actively burrowing species. These animals are often predators and their bodies more or less conform to the generalized polychaete structure. Some may be tube-dwellers but their tubes are not permanently fixed to the substratum.

Most of the species belonging to the Sedentaria live in tubes or burrows permanently situated in or on the substratum. The bodies of the sedentary polychaetes are greatly modified in structure, with the parapodia much reduced in size and have specialized structures developed for the collection of microscopic food particles. For example, in some sedentary tubiculous species, the head is provided with long tentacles that protrude from the mouth of the tube when the animal is feeding.

---

* Some authorities recognize the Archiannelida and Myzostomaria as separate classes, while others regard these animals as aberrant polychaetes. For the purpose of this work it is convenient to regard them as polychaetes.

These tentacles carry rows of cilia that waft microscopic plants and animals towards the mouth.

Polychaetes are found in all the oceans of the world. They are abundant in shallow coastal waters (especially in mud and sand) and in the intertidal zone, but they also occur at great depths. A small number of species are entirely pelagic, and others are pelagic only during the breeding season.

## Collecting

On rocky shores intertidal species may occasionally be found on the exposed surface of rocks, but they are much more commonly located in crevices, amongst seaweeds and encrusting organisms, or burrowing into rocks. Crevices can sometimes be prised open with a crowbar to reveal the worms. Weed and encrusting life should be removed from the rock and thoroughly washed in a bucket of sea water. Many of the bristleworms will be flushed from their hiding places with this method, but if time permits the material should be left for a few hours until the sea water becomes stale. Worms not dislodged by flushing will then leave their hiding places and collect on the sides and on the bottom of the container. They may also be induced to leave by adding fresh water or magnesium sulphate to the sea water. Finally, the material should be carefully teased apart to reveal any remaining worms. Rock-encrusting and rock-boring species can be collected by breaking off chunks of rock carrying or containing the polychaetes. These rock pieces should then be treated in the manner described above in order to induce the worms to leave their tubes or tunnels. The tubes should, as far as possible, be retained with the specimens.

Many species can be found by turning over stones on the shore. Species living in soft substrata such as sand, gravel or mud, can be collected by digging with a fork and sifting the substratum through the fingers to capture larger specimens and through a 1 mm sieve to retain smaller ones.

In calm weather some subtidal species can be collected by wading into the water below the low-tide mark; a glass-bottomed box placed on the water surface will eliminate ripples and allow an uninterrupted view of the bottom. In deeper water, polychaetes may be collected with reasonable success from soft substrata by using grabs and dredges (pp. 105–110), but diving techniques are necessary to collect successfully from a hard bottom (p. 103). Divers should carefully remove weeds and encrusting life which are likely to house polychaetes (the holdfasts of kelp are particularly productive) and place them in polythene bags. Small worms found in crevices and under stones can be picked up individually with blunt forceps and placed in polythene containers. Divers can sometimes persuade worms to leave crevices by squirting in commercial formalin solution from a squeeze-bottle. On returning to the surface much of the material can be examined in the manner described above for intertidal material.

Planktonic species can be attracted in large numbers by suspending a light in the water at night. The worms can then be captured with a net.

### Preserving

If the worms have to be dealt with quickly they can be placed directly in a fixative solution of 5 per cent buffered formalin or 10 per cent buffered Dowicil (p. 132) made up in sea water. Many polychaetes can be stored safely for several months in these solutions and indeed for delicate pelagic forms, these solutions should be used for permanent storage. Non-pelagic forms, however, are best transferred after 48 hours to 70–90 per cent alcohol or to 1 per cent propylene phenoxetol after washing in clean sea water.

With this rapid method of killing many of the more delicate worms may distort and contract, but this can be avoided if they are first anaes-thetized by the slow addition of alcohol to the water in which they are contained, or by the addition of a few crystals of MS 222 (p. 124) or magnesium sulphate. When the worms no longer respond to the touch they should be transferred to a flat dish containing the fixative solution (5 per cent formalin or 10 per cent Dowicil as noted above) and kept straight until they stiffen. Worms which have a protrusible proboscis should be made to extrude this by exerting pressure just behind the head.

Some authorities recommend Bouin's or Zenker's fixative (pp. 128, 130) if specimens are required for micro-anatomical work.

## Oligochaeta (*Earthworms, Potworms and Freshwater Ringed Worms*)

In the oligochaetes the parapodia are completely absent, but the lateral bristles are present in most species where they are implanted directly in the body wall. The head is also very much reduced and devoid of appendages. This class includes the familiar earthworms and allied freshwater species, as well as the semi-aquatic potworms (Enchytraei-dae). Oligochaetes vary tremendously in size. Some freshwater species are minute, scarcely exceeding a millimetre in length, whilst the giant Australian earthworms (*Megascolides* spp.) can attain lengths of several metres. The oligochaetes are cosmopolitan in distribution. Terrestrial forms are found in soil, amongst decaying plant material and in organic detritus of all kinds. Enchytraeidae, for example, often occur in very large numbers in sewage filters or rotting seaweed. The freshwater forms burrow in mud, silt and sand or live amongst submerged vegetation.

## Collecting

Earthworms are very easily transported from one part of the world to another, often in soil associated with plant material, and they quickly establish themselves in new areas. Even in relatively remote regions, worms occurring in gardens and other cultivated areas often prove to be common European species. For this reason searches for exotic indigenous species should be made in places remote from cultivation.

Earthworms can be obtained by hand-sorting samples of soil, moss, leaf litter and rotting vegetation of all kinds. Many soil-inhabiting species emerge from their burrows after rain, and at night (except during frost and bright moonlight) some species lie with the greater part of their bodies out of their holes. They are very sensitive to strong light and to sound vibrations, and if alarmed will retreat with surprising speed.

Chemicals are often used for extracting earthworms from soils. The materials most frequently used are proprietary brand vermifuges, potassium permanganate and formalin, but dilute solutions of mustard in water are also said to be effective. Earthworms can also be extracted by passing an alternating current of electricity through the soil. This method is sometimes employed for sampling earthworm populations on experimental plots that can neither be dug nor treated with chemical expellents.

Potworms can be collected by teasing apart soil and various types or organic material under a binocular microscope or lens. These animals can also be collected using the Baermann funnel (p. 95).

Freshwater oligochaetes can be collected by washing bottom sediments through a series of sieves. Specimens can also be obtained by rinsing the roots, stems and leaves of aquatic plants or by drawing a Birge net (p. 116) through the vegetation.

## Preserving

Before being killed by immersion in a fixative solution, live earthworms should be anaesthetized. Several anaesthetic solutions are effective, and 5–10 per cent alcohol or 1 per cent propylene phenoxetol are amongst the most convenient. After 10–15 minutes in the anaesthetic solution the specimens become limp and when they no longer respond to probing, can be transferred to a flat dish, straightened out, and killed by the addition of a small quantity of fixative solution. Material for general taxonomic studies can be fixed in 4 per cent formalin or 10 per cent Dowicil (p. 132) solutions. After 24 hours in the fixative they should be washed and stored in 70–90 per cent alcohol or in a 1 per cent solution of propylene phenoxetol.

The larger fresh-water oligochaetes can also be killed and fixed in the manner described for earthworms, but many workers advocate killing and fixing small oligochaetes without the use of an anaesthetic. Freshwater species (including Enchytraeidae) can be washed and placed in a shallow dish containing a little water. After a while they will begin to extend and can then be killed by adding a little concentrated formalin to the water. When collecting with a Birge net (p. 116), a little water should be poured off each time the container is changed, so that the worms can be killed by topping up with commercial formalin solution. After fixing for about 24 hours, aquatic worms should be preserved in 70–90 per cent alcohol, although they can be kept for long periods in formalin.


## Hirudinea (*Leeches*)

The leeches are cylindrical or flattened annelids. Most have no bristles and the body is annulated, although only a few of these external annulations correspond with the internal segmentation. The body is frequently tapered anteriorly and at each end segments are modified as suckers. The anterior sucker usually surrounds the mouth. The blood-sucking species are equipped with jaws for piercing the skin of the host.

While most are free-living predatory species, many leeches lead an ectoparasitic existence feeding on the blood of their vertebrate, and less frequently invertebrate, hosts. The majority of the blood-sucking leeches attach only temporarily to the host. However, some fish parasites belonging to the family Piscicolidae, and possibly a few species which are found in the nasal cavities of water-birds and mammals, live as adults more or less permanently attached to their host, except during the breeding periods. The blood suckers engorge at infrequent intervals and are able to survive long periods of fasting. They attach themselves first with the posterior sucker; after which the anterior sucker is applied to the skin and a wound is made, generally with the aid of toothed jaws. The digestive tract is then filled with blood and the leech drops off. During feeding the leech secretes an anticoagulant saliva (hirudin), and for this reason the wounds made by leeches tend to bleed for a long time.

Leeches occur in most parts of the world and are found on land and in water. Although there are a few marine species, the majority of aquatic leeches live in fresh water. The terrestrial species, which are primarily parasites of warm-blooded vertebrates, occur mainly in humid environments of tropical Asia.

### Collecting
Aquatic species can often be collected individually with forceps by turning over stones in shallow water and by sorting through debris. A dip-net can be used to capture rapidly swimming species. Unattached terrestrial species can also be picked up with forceps, and during the rainy season in tropical Asia they can be found with little difficulty. A quick spray with an aerosol insecticide will usually induce attached leeches to release their hold very quickly, but they can also be detached by dabbing them with alcohol, formalin or even salt. Species living in the nasal cavities of birds and mammals can be removed with forceps after spraying with a local anaesthetic.

### Preserving
Leeches must be relaxed with an anaesthetic before they are killed. From among the wide range of anaesthetic substances available, the most convenient for use in the field are a 5–10 per cent solution of alcohol or a 1 per cent solution of propylene phenoxetol. However, lemon juice may also be used and soda water is effective for small specimens. When the leeches are completely extended and no longer respond to probing, they should be transferred to a shallow dish, straightened out, and killed by pouring over them a 4 per cent solution of formalin. Small flattened forms can be extended by compressing them between two microscope slides held together with rubber bands. The specimens can then be killed by immersing the slides in the fixative. Leeches are best stored in 4 per cent formalin or in a 10 per cent solution of Dowicil, since alcohol and propylene phenoxetol tend to destroy their colour patterns.


# Pogonophora

These are slender, tentaculate marine worms in which the digestive tract is lacking. They are benthic animals that live in secreted tubes usually embedded in the bottom ooze. Most of the known species range from 100–350 mm in length, but their tubes are often much longer. Pogonophores are exclusively marine and occur mainly in deep water. The first detailed studies of their anatomy led to the view that their affinities lay with the Deuterostomia, in particular the Hemichordata (p. 84), but more recent studies have indicated that they may be related to the annelidan stock.

The geographical distribution of the Pogonophora is imperfectly known. Although specimens were first collected at the turn of the century, their true biological significance was not appreciated until relatively recently, and it seems likely that over the years a great deal of pogonophoran material in dredgings etc. was unrecognized and discarded. About one hundred species have been described. It is now known that they are a major element of the fauna on long stretches of the continental slope.

**Collecting**

Pogonophora can be taken by dredging, trawling or by the use of grabs and bottom corers although the specimens are usually damaged with their rear ends missing. In Scandinavian waters one species has been collected by divers at depths of only 15–20 m.

**Preserving**

The following notes are based on part of Ivanov's (1963) account of study methods.

The tubes of pogonophorans are rarely preserved intact, and because it is extremely difficult to extract most species from their tubes without damaging them, separate animals or parts of their bodies found in the contents of trawls and so on are particularly valuable.

Animals still in their tubes should, if the consistency of the tubes is sufficiently stiff, be fixed in 70 per cent alcohol. The alcohol should be changed after 24 hours. Species with soft tubes should be fixed in 2–3 per cent formalin. Material for histological study can be fixed in Bouin's solution (p. 128), Zenker's fluid (p. 130), or a saturated mercuric chloride solution with acetic acid (p. 131).

*Reference*
Ivanov, A. V. 1963. *Pogonophora* (English translation by D. B. Carlisle). New York: Academic Press.

# Arthropoda (*Jointed-limbed animals*) (Plate 8a–g, 9a–h)

As in the case of the annelids, the arthropod body is composed of a series of segments, but in the arthropods at least some and frequently nearly all of these segments carry a pair of jointed appendages. At least one pair of these appendages is used for feeding (jaws) and the remainder are modified for other purposes. Thus, some of the head appendages are sensory organs and the majority of those on the trunk function as locomotory organs. The arthropod body is covered by a thick cuticle that acts as an external skeleton. This cuticular skeleton is secreted by the animals (the principal component of one of its layers is a chemical called chitin) and to accommodate growth it is periodically moulted. The growth stage assumed between each moult (or ecdysis) is known as an instar.

Arthropods surpass all other animal groups both in diversity and in numbers of species. They are the only invertebrates that have been truly successful in exploiting all terrestrial habitats and they include somewhere in the region of 80 per cent of all the known animal species. Classification schemes for the Arthropoda are rather complex and for the purposes of this account the survivors of the phylum can be referred to one of six groups: Onychophora, Myriapoda-Insecta, Crustacea, Xiphosura, Arachnida, Pycnogonida. Of these, Xiphosura and Arachnida can be united within the Chelicerata. In this group the head appendages consist of a pair of feeding organs called chelicerae which, primitively, are pincer-like in appearance, and a pair of organs called pedipalps which are often sensory but can be variously modified. The Pycnogonida are also regarded as Chelicerata by some zoologists but these animals are extremely aberrant and their true affinities are uncertain. The Crustacea and Myriapoda-Insecta are often grouped together as the Mandibulata but this is a *grade* of structure rather than a natural group. These animals do not have chelicerae but possess, as head appendages, antennae (sometimes two pairs), mandibles and maxillae. It can also be noted that Manton (1973. *J. Zool. Lond.* **171,** 111–130) considers that the Arthropoda as a whole are probably a polyphyletic group and that 'arthropodization' has occurred at least three times forming the phyla Crustacea, Chelicerata and Uniramia (Onychophora, Myriapoda and Insecta).

The most successful of all the arthropods are the insects and the methods of collecting and preserving these animals are described in a separate Museum publication (*Instructions for Collectors No. 4A, Insects*).

# Onychophora

The Onychophora, comprising *Peripatus* and its allies, are terrestrial animals with a long caterpillar-like body. They have a single pair of antennae and numerous unsegmented stumpy legs bearing claws. Although the group includes only about 70 species, it is of great zoological interest as in a sense its members form a link between the soft-bodied annelid worms and the hard-skinned arthropods. The Onychophora have a thin cuticle, and they respire by means of a simple tracheal system with openings that cannot be closed. Consequently, these animals have little protection against water loss and they live only in humid habitats. They can be found for example amongst litter, under the bark of fallen trees and below stones in damp forests. One southern African species, *Peripatopsis alba*, is cavernicolous.

Onychophora occur in the tropics and in the temperate regions of the southern hemisphere. They are not found in temperate regions of the northern hemisphere except in certain parts of the eastern Himalayas, and in Africa do not occur north of the equator.

## Collecting
No special methods are employed. The animals can be picked from their hiding places with a pair of blunt forceps or manoeuvred into a tube. Great care should be taken not to overcollect as these animals cannot traverse dry areas to recolonise isolated humid habitats.

## Preserving
Onychophora can be killed with chloroform or ethyl acetate vapour or by means of an entomological killing-bottle. For gross morphological studies they can be fixed in either formalin or spirit, although they tend to shrink rather badly in the latter. If formalin is used, specimens should first be dipped into spirit in order to make the skin permeable, and then, if much shrunken, distended by soaking in water. They should finally be preserved in 5 per cent formalin solution or in 70–90 per cent alcohol. When animals are required for detailed anatomical studies they are best brought back alive to the laboratory and subjected to special fixation procedures (see Manton, 1937).

Plate 8 Crustacea and Pycnogonida
a, *Leionymphon*, a pycnogonid; b, *Daphnia*, a freshwater cladoceran; c, *Philoscia*, a terrestrial isopod, or woodlouse; d, *Portunus*, a swimming crab; e, *Calanus*, a planktonic marine copepod; f, *Balanus*, a sessile barnacle; g, *Diastylis*, a cumacean.

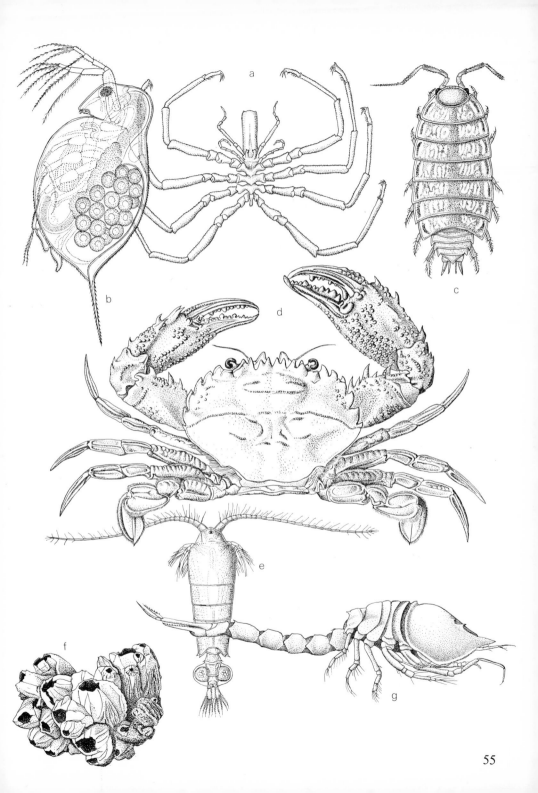

55

*Reference*
Manton, S. M. (1937). The feeding, digestion, excretion and food storage of *Peripatopsis*. *Phil. Trans. R. Soc.* B. **227**, 411–464.

# Myriapoda

For the purpose of this work the centipedes, millipedes and two less-familiar groups of terrestrial many-legged arthropods can be considered together as the Myriapoda, although this grouping is somewhat artificial. All the Myriapoda are many legged, terrestrial, mandibulate arthropods in which the head bears a single pair of antennae. There are four 'myriapodous' groups – Diplopoda (millipedes), Chilopoda (centipedes), Pauropoda and Symphyla.

In the millipedes the greater part of the trunk is made up of a large number of double segments, each double segment carrying two pairs of legs. The body in these animals is usually cylindrical, although there are a number of flat-backed species. Members of one order, Glomerida, the so-called pill millipedes, are able to roll themselves up into a ball. Millipedes are for the most part slow-moving species that feed predominantly on dead and decaying plant material although one or two species may feed on living plants. Their slow gait is adapted for exerting a powerful pushing force and enables the animals to push their way through soil and humus layers. They are cosmopolitan in distribution but reach their greatest abundance in tropical regions. Some of the tropical species belonging to the order Julida can attain lengths of nearly 300 mm.

In the centipedes the first trunk segment carries a pair of poison claws, and all the other trunk segments, except the last, carry a single pair of legs. The centipedes are primarily nocturnal predators, although a few species belonging to the order Geophilomorpha may on occasion feed on plant material. Centipedes are found under stones, in leaf litter, under bark and in crevices of the soil. A number of species are cave dwellers and a few are intertidal. They are cosmopolitan in distribution and many tropical and subtropical species are quite large, particularly those belonging to the order Scolopendromorpha.

The Pauropoda and Symphyla are small soft-bodied myriapods that

Plate 9 Ziphosura and Arachnida
a, *Tachypleus*, a king crab; b, *Centruroides*, a scorpion; c, a false scorpion; d, *Liobunum*, a harvestman; e, an amblypygid, or tail-less whip-scorpion; f, a solifugid, a camel-spider or sun-scorpion; g, *Araneus*, an orb-web spider; h, *Caloglyphus*, an astigmatid mite.

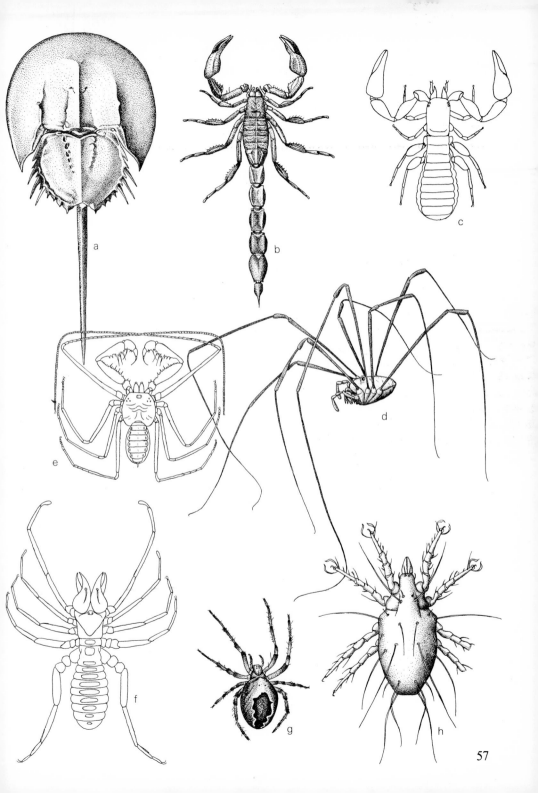

57

live in soil and leaf litter. Pauropods are slow-moving animals that probably feed mainly on fungi and dead plant material. They have curiously branched antennae. The Symphyla can run quite rapidly. They feed mainly on dead plant material although at least one species, *Scutigerella immaculata* – the so-called glasshouse centipede – attacks living plants. Pauropods and symphylids are widely distributed in tropical and temperate regions.

### Collecting
The larger forms can be picked up from their hiding places with blunt forceps and the smaller forms can often be captured with a moistened camel-hair brush.* Chilopods and diplopods can be obtained in potato and pitfall traps and the smaller forms of all groups can be collected by heat-desiccation methods (p. 92).

### Preserving
Myriapods can be killed and preserved in 70–90 per cent alcohol although it is often an advantage to kill or at least heavily anaesthetize the larger species with ethyl acetate vapour before immersing them in the preservative. The segments of very large millipedes tend to fall apart after some years in spirit and this may be the result of poor initial preservation and fixation. Consequently with these animals care should be taken to ensure that an adequate volume of spirit is used to allow for dilution due to dehydration, and it is advisable to transfer them to fresh spirit after 12–24 hours. Formalin should not be used for the preservation of myriapods.

## Crustacea (Plate 8b–g)

The crustaceans are mandibulate arthropods which are mainly aquatic and which breathe by means of gills or through the general surfaces of the body. Typically, they possess two pairs of antennae on the front part of the head and they have at least three pairs of mouthparts. Most crustaceans develop through one or more free-living aquatic larval

---

* Certain tropical and subtropical scolopendromorph centipedes should be handled cautiously – their bites can be painful. The symptoms evoked however are generally quite local and can usually be safely relieved with local anaesthetics of the type used by dentists. Large tropical millipedes should also be handled with care. Many species secrete caustic fluids and some can squirt the fluids over distances of a couple of metres.

stages, and for the group as a whole there is a bewildering variety of larval forms and accompanying larval terminology.

The Crustacea comprise eight subclasses, seven of which are frequently considered under the general term Entomostraca, although it must be noted that this is merely an expedient grouping and not a natural or phylogenetic assemblage. The entomostracan groups are Branchiopoda, Ostracoda, Copepoda, Branchiura, Cephalocarida, Mystacocarida, Cirripedia.

It is not possible to give a simple definition embracing all of the Entomostraca. They are generally very small or microscopic animals, and are extraordinarily diverse in form, displaying a very wide variety of adaptations to the habitats in which they are found. Several entomostracan groups can be legitimately called 'living fossils' for, judging from the evidence of immensely long fossil records, they appear to have survived with little change over very long periods of time.

The Branchiopoda comprises four orders of free-living, predominantly freshwater crustaceans – the Anostraca (fairy shrimps), Notostraca (tadpole shrimps) and Conchostraca (clam shrimps) all of which are characteristic inhabitants of temporary freshwater pools, usually in arid regions, and the Cladocera (water fleas) that are very abundant in freshwater habitats, but are also locally common in the inshore marine plankton. Typically, branchiopods are filter feeders although some cladoceran genera are predatory and have a much reduced carapace.

The Ostracoda (mussel shrimps) are small, marine and freshwater, crustaceans in which the body is entirely contained within a bivalved shell, rather resembling a small mussel. They are active swimmers occurring in both planktonic and benthic habitats, with different species exploiting almost all available food resources. In addition, they may occur in underground caves, or in damp terrestrial habitats such as moss and leaf litter.

Branchiura (fish lice) live as ectoparasites on marine and freshwater teleost fishes, and occasionally on amphibians. The body of the parasitic crustacean is strongly flattened and usually has powerful suckers for attachment, although they still retain the capacity for free-swimming away from the host.

The Cephalocarida and Mystacocarida are both very small groups of exclusively marine entomostracans. The mystacocarids are slender-bodied (up to 0·5 mm length), and live within the interstices of sandy inshore sediments of tropical and warm temperate seas. The primitive cephalocarids are much larger (2·0–4·0 mm length) and live as deposit feeders on the surface of fine sediments, in shallow and deep waters, or in association with the sediment around the roots of turtle grass.

The Cirripedia comprises the familiar barnacles found encrusting rocks, pilings, buoys, and other floating or fixed objects throughout the intertidal and shallow subtidal zones. In addition they can be found attached to other animals such as whales, turtles, or other large crustaceans. The cirripedes are exclusively marine, and are of considerable economic importance as fouling organisms, especially on the undersides of large ships such as tankers. Typically, the body of the barnacle is protected by a series of calcareous plates (acorn barnacles), but in some of the stalked barnacles (goose barnacles) the plates may be reduced or absent. A large number of cirripedes are parasitic, commonly on crabs, and have lost most of the obvious crustacean affinities, except for typical free-living larvae. The body is reduced to an external sac containing the reproductive organs, and an internal rootlet system. All cirripedes have retained the free-swimming planktonic larvae.

The Copepoda is by far the largest of the entomostracan subclasses with about 7500 recognized species. Copepods are abundant in almost all aquatic habitats, and the free-living calanoid and cyclopoid copepods comprise a significant proportion of marine and freshwater zooplankton. The harpacticoid copepods are predominantly marine and benthic, often interstitial. In addition, there is a very large number of parasitic or commensal copepod species. The caligoid copepods (sea lice), for example, are entirely parasitic, mostly on fishes, and may be so highly modified that their crustacean affinities are no longer obvious. Other copepod species can be found in associations with representatives of almost every major phylum of marine invertebrates.

In terms of the number of species, the Malacostraca is by far the largest subclass of the Crustacea. In addition to the familiar Decapoda (crabs, lobsters, crayfish, shrimps and prawns), Amphipoda (sandhoppers) and Isopoda (slaters, woodlice), the Malacostraca also contains a number of other less familiar shrimp-like groups (Stomatopoda (mantis shrimps), Mysidacea (opossum shrimps), Euphausiacea, Tanaidacea, Cumacea, Anaspidacea, Nebaliacea, Bathynellacea, Thermosbaenacea. Of these, the Thermosbaenacea which occur only in fresh or brackish subterranean waters and hot springs, and the Anaspidacea that are found only in certain caves and mountain pools, have a very restricted distribution, but others, such as the Mysidacea and Euphausiacea, are found in all the oceans of the world and the mysids also have representatives living in fresh and brackish water.

While there are some terrestrial Malacostraca (land crabs and woodlice), the members of this group are primarily aquatic and even the land-dwellers need acess to humid conditions or to bodies of water for respiration and reproduction. Some of the aquatic species are pelagic throughout their lives, but the majority live in or on a variety of substrata and many smaller species can be found attached to aquatic vegetation, on or within mud, sand and gravel, and some species are even capable of boring into timber and rock. Finally, several of the malacostracan groups contain species which are more-or-less modified for a parasitic existence. For example, the Isopoda has a large number of parasitic species, found attached to marine or fresh-water fish, other aquatic vertebrate, or parasitizing other crustaceans, especially decapods.

## Collecting

The smaller free-swimming forms, including the larval stages of many of the larger sessile or bottom-dwelling species, may be collected by the usual methods used for plankton (p. 110). Bottom-living marine forms can be collected intertidally by hand sampling or offshore shere the bottom is suitable, by dredge, trawl or grab (p. 105). Species living in situations inaccessible to the normal sampling techniques, for example amongst rocks and corals below the tidal zone, are more easily collected by diving (p. 103). Although the larger marine crustaceans are sometimes encountered in the open, they tend to be relatively re-tiring and should be looked for in rock crevices, coral formations, wooden piles and amongst weeds. The encrusting barnacles are best left attached to a piece of rock or other object, but if necessary they may be carefully prised off with a knife or similar instrument.

An effective way of collecting the bottom-living ostracods is to scrape off the swash-mark layer along a beach. This is the cuspate line left by each swash of the tide and, if looked at closely, it is seen to be dis-tinctly whiter than the surrounding beach sand. This whiteness is due to the presence of the shells and tests of some of the innumerable micro-scopic organisms which live in the littoral zone or offshore to a depth of about 15 metres. The tiny animals are simply floated in on each wave. Twenty minutes' careful scraping will generally yield enough shelly concentrate to fill a $\frac{1}{2}$ kg jam jar. Some beaches are more suitable for this purpose than others but collecting is invariably better in the aftermath of a storm. Once collected, the material is dried and then the microscopic organisms can be floated off by pouring the material into a larger beaker of carbon tetrachloride or dry-cleaning fluid. This should be done in the open air or in a fume cupboard since these volatile chemicals can have dangerous effects if inhaled. The 'float' is then filtered off and the chemical recovered for future use. The few grammes of 'float' almost invariably contain thousands of microscopic organ-isms. This technique does not destroy the soft parts of the ostracods and specimens for dissection collected in this way can be made pliable by soaking them in glycerol for a few hours.

Some of the larger crustaceans construct extensive burrows – in some cases a metre or more long – in soft substrata. These deep-burrowing creatures, for example some callianassids, may be difficult to dig out, but can often be enticed into the open by placing pieces of fresh crab meat near the burrow entrance. An interesting alternative method of bringing these animals to the surface is to drop small pieces of shell into the burrows, and with luck the inhabitant will come to the entrance to eject the debris. Below water, burrowing animals can be driven out by squirting a repellant such as a weak solution of formalin or domestic bleach into the openings of the burrow.

The larger species in relatively shallow water can be taken in traps similar to the well-known pots and kreels used by commercial crab and lobster fishermen. Another good way of collecting estuarine and near-shore marine crabs is to lower an old nylon stocking with some offal knotted in the toe into the water on a piece of string. The crabs' claws become entangled with the nylon and, provided a good spot is selected, several specimens can be collected in a relatively short time.

Many smaller crustaceans can be found amongst aquatic vegetation and should be searched for very carefully and thoroughly. In ponds and lakes some of them can be collected by dragging a hand-net, or a fine mesh tow-net protected by a core of coarse wire mesh, through the plant growth. Also the weed should be hand collected into a bowl or bucket and carefully sorted, if necessary under a binocular microscope. A small quantity of formalin solution or alcohol added to the water will often induce the animals to leave their hiding places and swim to the sides of the container. This method is particularly useful for isolating small entomostracans, such as copepods, which live in close association with other invertebrates.

Apart from the large and conspicuous land crabs, land-hermits and coconut crabs, the terrestrial Crustacea consists mainly of woodlice. These usually live in damp places – for example in the soil, under stones or amongst vegetation – although a few species inhabit sandy deserts. Small woodlice can be picked up with a moistened camel-hair brush. Woodlice can also be trapped with potato or pitfall traps, or collected from soil debris using a Berlese funnel (p. 92) or similar method.

Edible crustaceans should not be neglected on the assumption that they are well known. For example, prawns purchased in foreign markets are often of interest although details of the locality in which they were captured are rarely available.

**Preserving**

Before crustaceans are preserved, a note should be made of any colour markings since these patterns tend to disappear rapidly after preservation and can be valuable aids in identification.

The smaller crustaceans taken by tow-net, as well as the fairy shrimps, brine shrimps and water fleas, can be killed by adding a little concentrated formalin to the water. Once the animals are dead they should be fixed in fresh 5 per cent buffered formalin solution made up with sea water or fresh water as appropriate. Pelagic forms taken from deep water are often very soft, and are best inserted into a narrow tube containing 5 per cent formalin, to prevent the animal from curling up. Amphipods can be stored successfully in 5 per cent formalin, but most other groups should be transferred to 70–90 per cent alcohol after about 24 hours' fixation in 5 per cent formalin solution or 10 per cent Dowicil solution.

For general use with pelagic marine Crustacea, Steedman's solution (p. 127) of formalin, propylene phenoxetol and propylene glycol is strongly recommended. This mixture combines the fixative action of formaldehyde with the preservative properties of propylene phenoxetol and propylene glycol. Satisfactory results can be obtained with woodlice and other small isopods by dropping them directly into 70–90 per cent alcohol. Specimens should never be allowed to become entangled in cotton wool or other packing material.

Larger crustaceans such as crabs, lobsters, shrimps and prawns require special care in killing because they tend to cast appendages when exposed to chemicals like formalin, although limb shedding can usually be avoided if a weak (1–2 per cent) formalin solution is used. Alternatively, animals can be anaesthetized with chloral hydrate or Sandoz MS 222p. 124) before being killed. Most aquatic forms can also be killed by exposing them to a higher or lower temperature than that in which they normally live. Thus, cold water forms will soon die if which they normally live. Thus, cold water forms will soon die if exposed to the sun, and warm water species can be killed by chilling them with ice. Large crustaceans should be fixed for at least three to four days in 5 per cent buffered formalin. The addition of a little glycerol to the formalin will prevent the animals from becoming too brittle. After fixation, specimens should be washed with fresh water and transferred to 70 per cent spirit. To avoid loss of appendages during transit large specimens should be individually wrapped in cheese cloth or polythene and packed in rigid containers.

# Xiphosura (*King crabs or horse-shoe crabs*) (Plate 9a)

The Xiphosura are large marine chelicerates in which the anterior part of the body is covered dorsally by a large carapace. The prosoma is hinged to the opisthosoma (or abdomen) which bears a long tail-like spine. The prosoma carries six pairs of jointed appendages. The first pair, the chelicerae, are three-jointed, but the remaining prosomal appendages are six-jointed. In the females all these appendages terminate in rather slender pincer-like claws, but in the males the claws of the second pair (the pedipalps), and in some species those of the third pair, are thickened, and in all except one species, *Carcinoscorpius rotundicauda*, the fixed digit of these thickened claws is very much reduced so that the claws appear to be single. The penultimate segments of the last pair of legs carry a series of broad spines which can be spread out and pushed against soft sand or mud without sinking in. The basal segments of the prosomal appendages are all equipped with spiny processes (gnathobases), and the appendages are so arranged that the gnathobases surround the mouth. The animals feed principally on worms and molluscs and the food is crushed by the gnathobases as the appendages are moved. The appendages of the opisthosoma are greatly flattened, and all except the first pair, which form the genital operculum, support gill lamellae.

There are four living species of Xiphosura, and these animals are often referred to as 'living fossils' since in structure they closely resemble the earliest fossil representatives of the group. The genus *Limulus*, for example, has persisted for some 200 million years. King crabs live in shallow waters along sandy or muddy shores and all species habitually burrow into sand or mud in search of food. The four living species are classified in three genera. *Limulus polyphemus* is found on the eastern coast of North America from Nova Scotia to Yucatan. The genus *Tachypleus* includes two southeast Asian species, *T. gigas* and *T. tridentatus*, the latter species occurring as far north as the coast of Japan. *Carcinoscorpius rotundicaudata* is found in the Gulf of Bengal, Siam and along the Philippine coast.

## Collecting*

Most commonly these animals are collected by hand from intertidal areas, although larger sizes can be obtained from deeper water by means of traps, trawls or dredges (p. 105). Egg nests are a few centimetres beneath the surface within a wide band of the intertidal zone on pre-

---

* Notes on the collection of king crabs were kindly provided by Dr Carl N. Shuster, Jr.

dominantly sandy beaches. Often the site of these nests can be located by observing the feeding activity of shore birds at low tide and that of tidewater minnows during the high tides. Early growth stages are usually common in the immediate area of the nest and can be taken at low tide from intertidal flats. There the tiny animals make characteristic 3-rail tracks (by the outermost edges of the carapace and the telson) as they plough through the ooze. The older, hence larger, animals range further and further away from the nesting areas. Only during the spawning season do the adults come onto the beaches (along the Atlantic coast of the United States mating occurs during the full moon tides of early summer). The three Indo-Pacific species have similar life histories.

### Preserving

Large specimens can be injected with formalin solution (into the heart and ventrally into the tissues), and the digestive tract also flushed with formalin. These specimens can then be left to dry in the sun. Smaller specimens that can be bottled can be placed directly into 70–90 per cent alcohol, and for gross morphological studies no prior fixation is necessary, although it may be an advantage to anaesthetize the animals with propylene phenoxetol or MS 222 (pp. 124) before plunging them into spirit.


## Arachnida (Plate 9b–h)

With the exception of a small number of families which have become secondarily adapted to life in water, the arachnids are terrestrial chelicerates and in most groups the body is fairly clearly divided into two regions – an anterior prosoma or cephalothorax and a posterior opisthosoma or abdomen. However in one large subclass (the Acari) these body regions are not clearly differentiated. The appendages common to all arachnids comprise a pair of chelicerae, a pair of pedipalps and four pairs of legs. The chelicerae, and pedipalps can be variously modified.

This large class is made up of 11 subclasses – Scorpiones (scorpions), Pseudoscorpiones (false scorpions), Solifugae (sun spiders), Palpigradi, Schizopeltida, Thelyphonida (whip scorpions), Amblypygi, Araneae (spiders), Ricinulei, Opiliones (harvest spiders) and Acari (mites and ticks).

Members of the first ten subclasses are free-living predators, although some harvest spiders will also eat materials such as bread, fat, fungi, seeds and vegetable detritus. The Scorpiones, Solifugae, Thelyphonida and Amblypygi are relatively large arachnids that are found only in the warmer parts of the world. They are nocturnal and during the day hide in various situations, for example under stones, amongst detritus and in crevices in the soil. The Schizopeltida, which are mainly tropical in distribution, have similar habits but they are much smaller, measuring only 5–7 mm in length. Pseudoscorpiones, Palpigradi and Ricinulei are small arachnids (measuring up to about 10 mm in length) that are found mainly in soil or forest litter. The Palpigradi and Ricinulei are found only in warmer countries – indeed the Ricinulei are rather rare and known only from West Africa and from Central and South America, whilst the Pseudoscorpiones are cosmopolitan in distribution.

The familiar spiders (Araneae) and harvestmen (Opiliones) are found throughout the world and occur in a wide range of habitats. The spiders, with some 35 000 described species, are perhaps the best known of all arachnids, and the manner in which they use the silken threads produced by complex abdominal glands for various purposes is a feature of particular interest. Harvestmen superficially resemble spiders although in these animals the abdomen is joined to the cephalothorax along the whole of its breadth and not by a narrow pedicel as in spiders. The subclass includes life-forms ranging from short-legged sluggish species living in the extreme surface layers of soil to long-legged animals which are found amongst taller herbage.

In terms of numbers of species the Acari will probably prove to be the largest of all the arachnid subclasses. They rival the insects in their diversity of form and are unique amongst the arachnids in including parasitic and phytophagous species. The subclass is made up of seven orders – Notostigmata, Tetrastigmata, Mesostigmata, Metastigmata (ticks), Cryptostigmata (oribatid mites), Astigmata and Prostigmata.

The Notostigmata and Tetrastigmata, both made up of predatory species, have a rather restricted distribution in the warmer parts of the world, but all the other orders are cosmopolitan. The Mesostigmata are for the most part quick-moving predatory mites living mainly in soil and litter, but the order also includes some endo- and ectoparasites, and a number of species are important as vectors of disease organisms. The Metastigmata (ticks) are the largest acarines and all the species classified in this order are ectoparasites of terrestrial vertebrates. A number of ticks are important as pests of livestock and as vectors of human pathogens. The Cryptostigmata (oribatid mites) are small, darkly coloured, heavily sclerotized species resembling tiny beetles. They abound in soil and leaf litter, and apparently feed mainly on decaying plant material. Astigmata are small, lightly sclerotized species and this order includes fungus and detritus feeders as well as parasites of vertebrates. Most of the feather and fur mites belong to this group and some of its members are important as pests of stored food products. The Prostigmata is the most heterogeneous order of mites. It includes many free-living predatory forms and a number of its families are made up of species which are exclusively phytophagous. This order also includes parasites of vertebrates and invertebrates as well as a series of families (the Hydracarina) which have become secondarily adapted to an aquatic mode of life.

## Collecting

The larger arachnids, for example scorpions, mygalomorph spiders and Solifugae, can usually be lifted from their hiding places with blunt forceps or manoeuvred into tubes by prodding.* Medium-sized species, for example most spiders and harvestmen, and even some small species such as plant-feeding mites can sometimes be picked off their substratum with an aspirator (p. 91) or moistened camel-hair brush, but they are often more conveniently collected by sweeping and beating. Dead leaves and other loose material can be shaken over newspaper; low trees and shrubs can be beaten over a beating tray and low herbage can be swept with a canvas net. An aspirator or moistened brush can then be used for transferring the material to the preservative. Many medium-sized arachnids can also be taken in traps of various sorts.

The small free-living arachnids, notably mites and pseudoscorpions, that abound in soil and in all sorts of organic detritus, are generally collected by means of heat-desiccation funnels (p. 92) or exceptionally by means of flotation techniques (p. 95). Water mites can be collected by sweeping water-weeds with a hand-net or by tow-netting. They should also be searched for by washing and sieving bunches of weed and by examining growth on submerged stones.

Ticks and other parasitic Acari should be collected by the methods outlined on pp. 116–118. Ticks are usually large enough to be easily seen, but on removal care should be taken not to leave their mouthparts buried in the host. If the tick is firmly attached to the host, the skin near the point of attachment should be dabbed with a little spirit and a few seconds allowed to elapse before pulling it off. Parasitic mites

---

* Although the majority of the Arachnida are quite harmless, some spiders and scorpions are dangerously venomous. Toxicity is not related to size and indeed some species with very potent venom are medium-sized forms. Before working in habitats where dangerous species are likely to be encountered, collectors are advised to familiarize themselves with the methods of treating bites and stings, and the appropriate equipment should be carried on collecting trips. Modern methods of treatment are described by Stahnke (1966).

may be very difficult to detect and a close examination of the body of the host with a lens is often necessary. If the bodies of the hosts are to be transported before examination, they should be tied up separately in polythene bags in order to prevent the escape of parasites and contamination by parasites from other bodies. It is important to note the name of the host, the site of attachment of parasites and the approximate degree of infestation.

**Preserving**
Most arachnids can be killed and preserved in 70–90 per cent alcohol. When the material is required for general taxonomic work no special fixation procedures are necessary. However, with the larger species, and even with many medium-sized spiders, it is often an advantage to kill or at least heavily anaesthetize the animals with ethyl acetate vapour before immersing them in the preservative, since when killed in spirit they frequently become set with their appendages firmly contracted to the body. Specimens treated with ethyl acetate can be set with their appendages extended as soon as they become limp. If the best results are to be obtained other methods should also be used for the gall mites (Eriophyoidea) and for the water mites (Hydracarina). Plant material infested with gall mites should be wrapped in soft tissue paper and allowed to dry. After drying, the material can be stored indefinitely in bags or envelopes and the mites recovered at any time by a fairly simple laboratory procedure (see Evans *et al.*, 1961). Water mites should be killed and stored in Viets' solution (p. 129). For terrestrial mites many workers favour Oudemans' fluid (p. 129) as a preservative, but with the newer methods of preparing mites for microscopic study, the advantages of this fluid are at best marginal. Formalin should never be used as a preservative for arachnids since it makes material brittle and very difficult to handle.

During recent years propylene phenoxetol (p. 135) has been used successfully as a post-fixation preservative for spiders. Cooke (1969) recommends that the living material should be collected into a 1–2 per cent aqueous solution of propylene phenoxetol, because in this solution spiders die with their limbs extended and much more rapidly than they would in alcohol. The catch is then transferred to 70 per cent alcohol and left overnight for fixation before being put back into propylene phenoxetol for storage.

*References*
Cooke, J. A. L. 1969. Notes on some useful arachnological techniques. *Bull. Brit. arach. Soc.* **1,** 42–43.

Evans, G. O., Sheals, J. G., & Macfarlane, D. 1961. *The Terrestrial Acari of the British Isles.* I. *Introduction and Biology.* London: British Museum (Natural History). 219 pp.

Stahnke, H. L. 1966. *The Treatment of Venomous Bites and Stings.* Tempe, Arizona: Arizona State University. 117 pp.

## Pycnogonida (Sea spiders) (Plate 8a)

The Pycnogonida or Pantopoda comprise about 600 species of marine spider-like animals. The narrow body is usually rod-shaped and consists almost entirely of the appendage-bearing prosoma, with the opisthosoma or abdomen being reduced to a minute stump. The digestive system has diverticula which extend into all or some of the appendages. The appendages number between four and nine pairs, and the first three pairs, which are not used for walking, are the chelicerae, pedipalps and ovigerous legs. In the female these first three pairs of appendages may become vestigial during postembryonic development, but in the male only the first two pairs may become reduced. The ovigerous legs are used by the male to carry the developing eggs. The posterior appendages are similar to each other and function as walking legs.

Pycnogonids occur from the intertidal regions to depths of nearly 7000 m. They feed on soft-bodied animals, especially on hydroid polyps and on medusae. They are predatory on small animals, but they tend to feed as external parasites on larger marine organisms. Pycnogonids occur in all the oceans of the world but they attain greatest abundance in colder waters.

### Collecting
Pycnogonids can be obtained by the ordinary methods of shore collecting (p. 96), diving (p. 103), and from the sea-bed by means of trawls, dredges and grabs (p. 105).

### Preserving
Material for general taxonomic use can be placed directly into 70–90 per cent alcohol.

## Tardigrada (*Bear-animalcules or water-bears*)

These are small arthropod-like animals with four pairs of stumpy legs each armed with either four separate claws or two pairs of double claws.

The adults range in length from about 0·05 to 1·2 mm but the majority are less than 0·5 mm long. Although strictly aquatic animals, tardigrades are most commonly found in films or droplets of water on terrestrial plants, for example mosses (especially those growing on moist trees, walls and in gutters), liverworts, lichens and certain angiosperms with a rosette growth form. However, they also occur on aquatic mosses and algae, on rooted freshwater plants and in the mud, sand and debris of ponds and lakes. About 25 species are marine. These occur amongst sand grains in shallow water, on pilings, stones and amongst algae. One species, *Tetrakenon synaptae*, lives amongst the tentacles of the Atlantic sea cucumber, *Leptosynapta inhaerans*, and another, *Pleocola limnoriae*, lives in close association with the isopod *Limnoria lignorum*.

Tardigrades are world-wide in distribution. About 350 species have been described and the great majority appear to feed on the contents of plant cells which they pierce by means of mouth stylets. Many species, particularly those associated with terrestrial plants, are able to survive long periods of adverse conditions (for example aridity, heat or extreme cold) by encystment or by passing into a shrivelled anabiotic state. These quiescent periods may last for several years and may be entered into several times during the course of the animal's life.

**Collecting**
Tardigrada may be collected from wet mosses, filamentous algae and so on by simply rinsing the plant material and examining the washings under a binocular microscope. The animals can be picked up and transferred to a preservative solution using a fine pipette. Dried plant material (for example mosses and lichens) should be soaked in water for several hours to several days and rinsed at intervals. Tardigrades may then be found in the washings as they emerge from their quiescent phase. Marine and freshwater species inhabiting mud and sand can be obtained by the sieving and flotation methods used for collecting the components of the benthic meiofauna (p. 98). Tradigrades are often found in considerable numbers amongst collections of soil micro-arthropods obtained by using flotation techniques (p. 95). They are also sometimes found amongst material collected from organic debris with the Baermann funnel technique (p. 95).

**Preserving**
Tardigrada can be killed, fixed and preserved in 5 per cent formalin solution or in 70–80 per cent alcohol.

## Pentastomida (*Tongue worms*)

The Pentastomida (known also as the Linguatulida) are blood-sucking endoparasites that, as adults, live in the nose, nasal sinuses or other parts of the respiratory tract of carnivorous terrestrial vertebrates. The principal hosts are tropical reptiles such as snakes and crocodiles, but some species parasitize birds and mammals. With only one known exception, the larvae are found (often encysted) in the viscera of an intermediate vertebrate host, and the intermediate hosts are generally herbivorous animals that fall prey to the hosts of the adult parasites.* The body of the adult, which can range in length from about 5 to 130 mm, is elongate and more or less clearly annulated. The newly hatched larva has 2 pairs of well-developed limbs each with a pair of terminal claws, but during development the limbs become reduced leaving only the hooked claws on the head region. The sexes are separate, males being much smaller than females.

The group is divided into two orders – Cephalobaenida and Porocephalida. The adults of the Cephalobaenida occur in reptiles and birds, and their principal intermediate hosts are reptiles and amphibia. The Porocephalida are made up of two superfamilies – Porocephaloidea and Linguatuloidea. The Porocephaloidea are, as adults, parasites of reptiles, and as larvae occur in fish, reptiles, mammals but rarely in birds. The Linguatuloidea are essentially parasites of mammals, both as adults and as larvae.

### Collecting
Although adults are sometimes found in nasal secretions, Pentastomida are generally collected by dissecting the hosts.

### Preserving
Pentastomida can be fixed and stored in 3–5 per cent formalin or 70–90 per cent alcohol.

## Mollusca (Plate 10a–e)

The Mollusca constitute one of the major divisions of the animal kingdom. There are probably over 100 000 living species and like

---

* In Singapore the larvae of *Raillietiella hemidactyli* Hett have recently been found in the cockroach (*Periplaneta americana*). The adults of this pentastomid parasitize geckonid lizards.

members of other successful phyla, these animals have adapted themselves to practically every available environmental condition. They are found in the sea, at all depths from the intertidal zones to the abyssal trenches, in all freshwater habitats, and on land to altitudes as high as the permanent snowline. It is not possible to define the phylum in terms of a small number of characters exhibited by every mollusc – although the basic body plan and particularly the developmental plan of all molluscs is similar, they can only be separated from other phyla by reference to a series of features. This series includes the presence of a clearly defined muscular foot, head and visceral mass. The visceral mass is usually enclosed to a varying degree by a mantle which, typically, is a flap of tissue that also encloses a cavity associated with respiration. Exterior to this is a shell, often brightly coloured and ornamented, into which the whole animal may retract. However, in a large number of species the shell is very much reduced or completely absent. In a number of molluscan groups the mouth leads into a relatively large muscular chamber in which is found a peculiar feeding organ known as the radula. This is a horny ribbon covered with recurved teeth. A complicated array of muscles pulls the radula back and forth over a cartilaginous base, and during feeding it can be protruded through the mouth. Not all molluscan classes possess a radula; nevertheless it is characteristically a molluscan organ for no similar structure is found in any other animal group.

The molluscs living at present are grouped in six classes – Monoplacophora, Aplacophora, Polyplacophora ( = Loricata), Scaphopoda, Gastropoda, Bivalvia ( = Lamellibranchia) and Cephalopoda.

A single genus, *Neopilina*, is the only living representative of the Monoplacophora, a class that at one time was extremely numerous. These animals are interesting as examples of molluscs that may possibly exhibit some form of metameric segmentation. The four recorded species are small with a conical shell having the apex centrally or anteriorly placed. The ventral surface is dominated by the foot and the surrounding pallial groove which contains five or six pairs of gills. This groove separates the edge of the foot from the enveloping mantle and the shell. The mouth, surrounded by various palps, lies in front of the foot. Representatives of the genus have been dredged from deep trenches in the Atlantic, Pacific and Indian Oceans.


Plate 10 Mollusca
a, *Coryphella*, a nudibranch; b, *Tonicella*, a chiton; c, *Paroctopus*, an octopus; d, *Buccinum*, a gastropod mollusc; e, *Cardium*, a bivalve or lamellibranch.

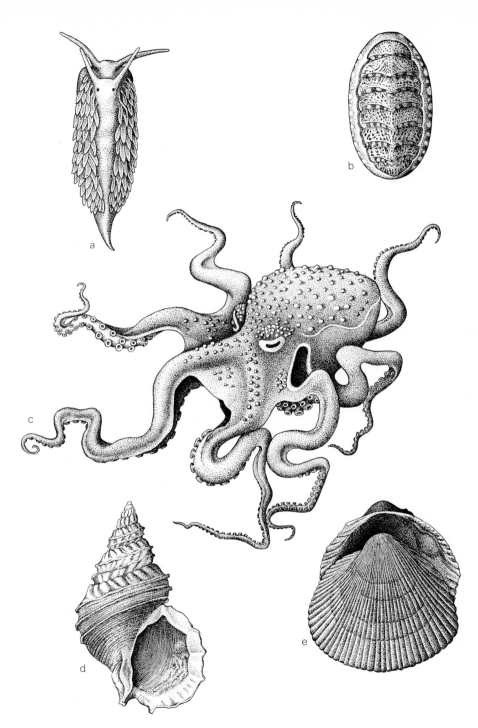

a

b

c

d

e

73

The class Aplacophora includes a number of worm-like marine molluscs that are found either living freely in soft substrata or in close association with hydroids. They have no shell. It is frequently difficult to distinguish any external features which indicate that these animals are in fact molluscs, and for this reason they are probably often overlooked. They occur mainly in deep water and their distribution is worldwide.

The class Polyplacophora or Loricata comprises the familiar chitons. These are all marine animals that are usually associated with hard substrata in the littoral region, although some species have been found at depths of about 750 m. Their distribution is world-wide. Chitons are bilaterally symmetrical and dorsoventrally flattened. The characteristic feature of this group is the dorsal shell made up of eight overlapping plates. They have a prominent muscular foot on the ventral surface and a well-developed radula.

The class Scaphopoda is made up of about 200 species of burrowing marine molluscs that are commonly known as elephant-tusk shells. They have a greatly elongated body and their tube-like shells are open at both ends. The mantle cavity is large and extends the entire length of the ventral surface. Scaphopods have a world-wide distribution and are found on soft substrata from shallow depths to the abyssal regions. Frequently they are found buried with only the small posterior aperture of the shell protruding into the water. They feed on microscopic organisms in the sand and surrounding water.

The class Gastropoda, which includes such common animals as snails, slugs, limpets and whelks, is by far the largest class of the Mollusca. Most of these animals have a well-defined head with eyes and tentacles, a large fleshy foot, and, typically, possess a spirally coiled shell into which the animal can retract. In slug forms, however, the shell is very much reduced or even completely absent, but when present it can be internal or external. Gastropods have a large visceral mass and a mantle which usually delimits a cavity that occupies the final whorl of the shell. Most species possess a radula, although in some, for example species of the genus *Conus*, this organ is greatly modified. The class has undergone very extensive adaptive radiation, and consequently its component groups exhibit great diversity of form. Marine species have become adapted to a benthic existence on or in all types of substrata as well as to a pelagic life, but there are also a number of boring and parasitic species. The group has invaded fresh water, and the pulmonate snails and slugs plus a few groups of operculate gastropods have colonized land habitats, eliminating gills and converting part of their mantle cavity into a 'lung' for air-breathing. The feeding habits of the group are variable. Many species are grazers, rasping off plant material with their radula. A number, however, are active predators and some are parasites.

Representatives of the class Bivalvia or Lamellibranchia occur in both marine and freshwater habitats and the class includes such common molluscs as cockles, mussels and oysters. Bivalves are all laterally compressed and they are characterized by the presence of a shell made up of two valves hinged together by an elastic proteinaceous ligament. They do not possess a head, buccal mass or radula. The two valves of the shell are pulled together by two large muscles called adductors which act antagonistically to the elastic hinge ligament. Inside the shell the two flaps of the mantle enclose a cavity occupied by the ciliated gills that sift particles of food from the current of water flowing through the cavity. On either side the mantle is attached by muscles to the inside of the shell along a semi-circular line a short distance from the shell edge. However, despite this attachment a foreign body (often a parasite) is occasionally lodged between the mantle and the shell. This foreign body then becomes the nucleus around which concentric layers of nacreous shell are laid down, and in this way a pearl is formed.

Bivalves are primarily inhabitants of sandy or muddy bottoms but many have become adapted for life on harder substrata. These forms have become completely sessile being either cemented directly to the substratum or attached to it by byssus threads. A number of species are borers, and while many of these confine their activities to softer clay banks some are capable of piercing very hard rocks. One of the most familiar of the boring bivalves is the highly specialized ship worm, *Teredo navalis* that causes extensive damage to wooden-hulled vessels and wooden pilings.

The cephalopods, which comprise the squids, cuttlefish, octopods and *Nautilus*, are the most highly organized of all the Mollusca. All are marine predators, and while a few species have adopted a less active, more or less benthic life, the majority are pelagic, swimming very rapidly by a system of jet propulsion. The name Cephalopoda ('head-footed') refers to the continuity of the head with the surrounding tentacles. Immediately behind the tentacles, on either side, is a well-developed eye, followed by the visceral mass surrounded by a thick muscular mantle. Anteriorly (i.e. towards the head-foot end) the mantle has a free edge (the collar) which can be closely pressed on to the visceral mass and the ventral funnel that leads from the mantle cavity to open under the head. When the mantle is relaxed water enters the mantle cavity around the edge of the collar, but when it contracts the free edges are sealed and a jet of water is forced out through the ventral funnel.

A completely developed external shell is found only in species of *Nautilus*. In the cuttlefish and squids the shell is reduced and internal while in the octopods it is completely lacking. *Nautilus* species can be found in relatively shallow water (about 35 m) around coral reefs in the south-western Pacific, usually swimming near the bottom. Many other cephalopods are much more widely distributed. Although some octopods for example *Eledonella* species, live at great depths, the majority occur mainly in shallow water on rocky bottoms. When not in search of food they can be found hiding in crevices and sometimes in stone dens of their own construction. Cuttlefish (*Sepia* spp.) live chiefly on sandy bottoms, but they are often found suspended, neutrally buoyant, in mid-water. Squids generally inhabit the surface water, although some, including the giant squids (*Architeuthis* spp.), which are by far the largest invertebrates, probably live at greater depths and only occasionally ascend to the surface.

### Collecting

Terrestrial gastropods can be found in a wide variety of situations, and amongst the more productive sites are crevices of rocks, under stones, in forest litter, amongst fallen wood, on tree trunks, under bark and even in the upper canopies of trees. In the wet tropics, in particular, examination of the latter can be very rewarding. The larger species can be collected by hand-sorting debris of various sorts and by manual searches in other suitable habitats. Smaller species living on herbage may be collected by sweeping and beating techniques (p. 91). Small litter-inhabiting snails can be collected by dry-sieving the litter through a series of graded sieves, each size group being hand-sorted for small snails. However, rather better catches of litter-dwelling forms can be expected with Williamson's wet-sieving method (1959) or with the flotation techniques devised by Salt & Hollick (1944), South (1964). When terrestrial gastropod activity is high, normally during relatively warm and humid conditions at night, very good results can be obtained with careful searching for active animals by torchlight or with chemical attractants. Many of the more effective baits utilize metaldehyde as one of the active constituents; however, the use of these chemicals can lead to gross distortion of the soft parts.

Alkaline waters with plenty of water-weeds are generally the richest situations for collecting freshwater species. In shallow streams and ponds a coarse hand-net or a wire sieve on a long pole can be used to collect the weeds, stones, mud and gravel, while in deeper water material can be collected with a dredge or plant grapple. Material obtained in this way can be hand-sorted or subjected to various washing and sieving techniques.

Intertidal marine molluscs can be obtained by the usual methods of shore collecting. Rocks harbouring burrowing and sessile species may have to be split with a hammer and cold chisel or even prised apart with a crowbar, but most species living in soft substrata can be fairly easily obtained by digging and sifting. The minute interstitial forms can be isolated by the specialized procedures applied to this fauna (p. 98). The inhabitants of deeper waters can be obtained by dredging, trawling and tow-netting although, as in the case of other marine groups, benthic species living on hard bottoms in moderately deep water are best collected by free diving (p. 103). It is worth noting that certain commensal and carnivorous species are frequently found in close association with various coelenterates and echinoderms. Pots baited with rotting fish or crabs will attract various scavenging and carnivorous species, and it is worthwhile examining the gut contents of fish, as these have been an important source of the rarer deep-sea forms, for example some species of cowries.

Planktonic larval forms are collected by means of fine tow-nets, while the young or juvenile animals (spat) that have settled onto the substratum can be obtained in a similar manner to the adults.

**Preserving**
Most molluscs contract violently when killed and have to be anaesthetized before fixation. Although a wide range of procedures and chemicals has been suggested for this operation the particular virtues of individual methods have frequently not been evaluated; a valuable resume of these and other preservation techniques is provided in Steedman (1976). Most terrestrial gastropods can be relaxed by immersing them until moribund or dead in a sealed container of cold water that has been deprived of its air by boiling. The animals may have to be left in the water for 24 hours or even longer before they are completely relaxed, but the process can be accelerated by adding anaesthetic substances such as urethane or menthol. Most of the aquatic molluscs can be anaesthetized with magnesium sulphate, magnesium chloride, urethane or menthol and with bivalve species good results have been obtained with propylene phenoxetol and phenoxetol BPC. It should be noted that animals left for too long in aqueous solutions tend to decompose very rapidly. If the necessary facilities are available chitons and some gastropods such as the marine slugs (order Nudibranchia) can be relaxed by freezing. The animals should be placed in a small quantity of sea water and left to extend. The vessel is then placed in an ice

chest or refrigerator and left until the water is cold, after which it is transferred to a freezer and the water containing the animals frozen solid. Thereafter surplus ice should be chipped off and the ice-embedded animals immersed in strong formalin. As the ice melts the animals are gradually exposed and fixed. For general taxonomic purposes the entire animal of most large marine molluscs can be fixed and preserved in 5 per cent neutral formalin, while 70 per cent alcohol is generally used as a fixative and preservative for terrestrial and freshwater species. For the long-term preservation of well-fixed material propylene phenoxetol has been found to be satisfactory for many molluscan groups.

It may be necessary to use certain specialized techniques for the fixation of more delicate species and specimens to be used for histological studies. Certainly, it would be advantageous to break the shell or slit the soft parts of at least some specimens in a collection so that the fixative can penetrate to all areas. The marked changes that occur in the pH of some formaldehyde-based fixatives during the early stages of fixation can result in the destruction of the detailed surface structure of the shell, although it is possible to reduce this effect by careful monitoring and buffering of the pH of the fluid during the initial stages of preservation (see Turner, R. D., in Steedman, 1976). However, these operations are time-consuming and it is often advisable to preserve a range of specimens using a variety of different methods, including the retention of some dry specimens. Small molluscs can be air dried, but for larger specimens it is sometimes necessary to remove the soft parts. This can be achieved by immersing the whole animal in boiling water and then using a bent pin or needle to extract the contents of the shell; the length of time the animal is immersed is critical and can only be learnt by experience. If small parts of the animal remain in the upper whorls these may be removed by spraying a fine jet of water into the shell aperture. Other methods of extracting the soft parts have been used successfully including, for example, deep freezing, immersing in paraffin or burying in ants' nests.

*References*

Salt, G., & Hollick, F. S. J. 1944. Studies of wireworm populations: I. A census of wireworms in pasture. *Ann. appl. Biol.* **31,** 53–64.

South A. 1964. Estimation of slug populations. *Ann. appl. Biol.* **53,** 251–258.

Steedman, H. F. 1976. *Zooplankton Fixation and Preservation.* Paris: The Unesco Press. 350 pp.

Williamson, M. H. 1959. The separation of molluscs from woodland leaf litter. *J. anim. Ecol.* **28,** 153–155.

# Chaetognatha (*Arrow-worms*) (Plate 12d)

The chaetognaths are small coelomate bilaterally-symmetrical, marine animals. The phylum contains about 50 known species. The torpedo-shaped body has a distinct rounded head which bears a pair of eyes and a number of large curved spines that are used for seizing the prey. Posteriorly the trunk is furnished with one or two pairs of lateral fins and a single caudal fin. These fins give the animals their characteristic dart-like appearance. Most arrow–worms are in the region of 30 mm in length although a few can attain a length of about 100 mm. With the exception of species of the genus *Spadella*, which are benthic in shallow waters, the chaetognaths are planktonic and indeed they are amongst the most common animals found in marine plankton. All are voracious predators and they will consume almost any suitably sized animal with which they come into contact. Although they are found in plankton samples from all oceans, arrow-worms attain their greatest abundance in tropical and sub-tropical waters.

### Collecting
Chaetognaths are collected by means of tow-nets and other methods employed for marine plankton (p. 110).

### Preserving
Arrow-worms can be fixed and stored in 5 per cent formalin solution or 70–90 per cent alcohol.

# Echinodermata (Plate 11a–f)

One of the most striking features of these relatively large marine invertebrates is their radial symmetry. In all the component groups of the phylum – except the sea cucumbers and certain echinoids where the design is obscured – the body is clearly made up of a number of similar parts (normally five) arranged radially around a central axis. However, this pentaradiate symmetry has been secondarily derived from a bilateral symmetry and is quite different from the primary radial symmetry found in the sponges, coelenterates and ctenophores to which the echinoderms are in no way related. Echinoderms have bilaterally symmetrical larvae that in many respects resemble the larvae of certain primitive chordates, and the structural organization of the body in the adults is at quite a high level. They have for example a nervous system and a well-developed digestive tract suspended in an extensive body cavity (coelom). Another striking and indeed unique feature of star-fishes is their method of locomotion. They move by means of numerous hollow extensible tentacle-like structures called podia or tube-feet located in (ambulacral) tracts corresponding to the rays, and these are activated by a peculiar hydraulic water vascular system. In most other echinoderms the tube-feet are used more in feeding than in locomotion, which is achieved by using the spines as stilts (in many echinoids), by arm flexure (in ophiuroids and crinoids) or by muscular contractions (in holothurians). Echinoderms have a skeleton composed of cal-careous plates or rods embedded in the body wall and typically this skeleton bears a number of projections which give the animals a spiny appearance, hence Echinodermata ('spiny skinned').

The phylum is made up of five classes – Asteroidea (starfishes), Ophiuroidea (brittle-stars and basket-stars), Echinoidea (sea-urchins, heart-urchins and cake-urchins or sand-dollars), Holothurioidea (sea-cucumbers), and Crinoidea (feather-stars and sea-lilies).

In the asteroids the body is made up of a number of arms radiating from and more or less merging with a central disc. Most starfishes have five arms but a greater number is found in some species. The sunstars (*Solaster* and *Heliaster* for example) may have up to 50.

Members of the Ophiuroidea are characterized by their small rounded central discs and long slender spiny arms. In some, these arms tend to break quite easily when handled, hence the common name 'brittle-stars'. Typically these animals have five arms. In some species (the basket-stars) the arms are branched repeatedly either from the base or more distally, often producing a gorgon-like mass of curling tendrils.

The Echinoidea do not have arms and the skeletal elements in their body walls interlock to form a rigid shell or test. Their bodies are sub-spherical (sea-urchins), flattened (cake-urchins or sand-dollars) or ovate (heart-urchins). The sea-urchins are often referred to as regular echinoids for in these animals the radial symmetry is preserved. In the irregular echinoids a secondary bilateral symmetry is superimposed on the radial plan.

Like the echinoids the body of the Holothurioidea is not drawn out into arms, but in this class the skeletal elements in the body wall are very much reduced so that the body surface is usually leathery and

Plate 11 Echinodermata

a, *Marthasterias*, a spiny starfish; b, *Amphiura*, an ophiuroid or brittle-star; c, *Monachometra*, a crinoid or feather-star; d, *Echinocardium*, a burrowing echinoid, or heart-urchin; e, *Paracentrotus*, an echinoid with erect spines; f, *Holothuria*, a sea-cucumber.

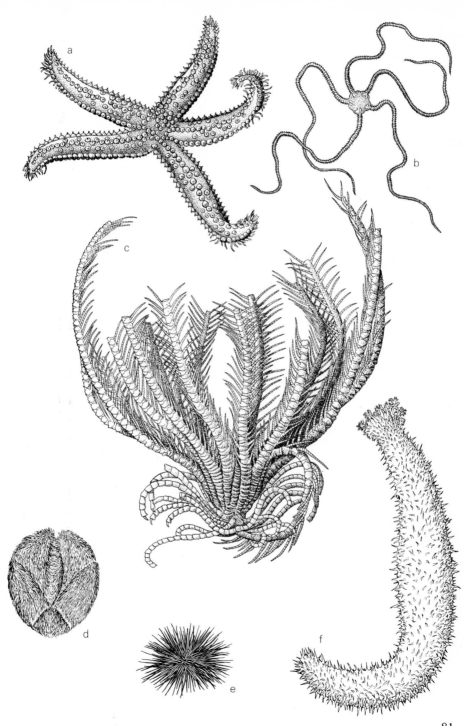

flexible. In the Holothurioidea the main axis of the body is elongate and has become horizontal but the radial symmetry is somewhat obscured by secondary dorso-ventral and/or antero-posterior specialization. Another distinguishing feature of the sea-cucumbers is that the tube-feet in the vicinity of the mouth are modified to form a ring of tentacles.

The Crinoidea includes both sessile (sea-lilies) and free-moving species (feather-stars). The rare, deep water sessile sea-lilies have a well-developed stalk, which in some species can be up to 600 mm long, and the basal part of the stalk carries a flattened disc or a system of root-like extensions which attaches the animal to the substratum. The stalk may also carry whorls of slender appendages called cirri. In the free-moving feather-stars the stalk is shed below the top-most joint or cen-trodorsal which usually bears cirri for temporary attachment. As in the asteroids and ophiuroids the body extends into five arms. In most species these divide basically only once to produce 10 arms, but some may possess up to 200 arms formed by repeated division. In the crinoids the arms have a jointed appearance and each arm carries two lateral rows of appendages called pinnules which give the animals a feathery appearance.

The echinoderms are exclusively marine animals and apart from a few pelagic holothurians their habits are benthic. They are found in all the oceans of the world from the intertidal zones to the greatest depths. Their distribution is often markedly aggregated, and the aggregations can often be very dense. This is particularly true of some brittle-stars and feather-stars, which are often found in patches of enormous numbers on the ocean floor. A few representatives occur above the low-water mark amongst weeds, under stones, in rock pools and amongst coral formations, but the vast majority are sublittoral. All classes except the Crinoidea include a number of species that live wholly or partly buried in muddy or sandy bottoms, and a few sea-urchins bore hollows in rocks. Deep-water forms are known from all groups and the stalked crinoids in particular tend to occur mainly at great depths.

**Collecting**
The intertidal species can be collected by the usual methods of shore collecting (p. 96) and the burrowing forms such as *Astropecten* (Asteroidea), *Echinocardium* (Echinoidea), *Amphiura* (Ophiuroidea) and *Cucumaria* (Holothurioidea) can often be obtained by digging in sand, mud or shingle at low-water level. Special care must be taken in handling feather-stars and many brittle-stars as they may fragment if roughly treated. For the more prickly sea-urchins it is advisable to wear protective gloves. A few tropical sea-urchins have irritant secretions in the

skin covering the spines or even small poison glands, but most can be handled safely with care.

The deeper-water benthic echinoderms can be collected by means of trawls, dredges and grabs (pp. 105–110), whilst shallow littoral species can be sampled very effectively by diving (p. 103).

**Preserving**

Most hard-bodied echinoderms can be placed directly in a fixative solution, but soft-bodied forms, and most especially holothurians, must be anaesthetized. Brittle-stars and feather-stars which tend to fragment rather easily are also generally better for some treatment prior to fixation – immersion in fresh water for a few hours, or refrigeration if available, is often sufficient to stun them. Holothurians can be left to expand and extend their tentacles in a container of sea water, and then a few drops of 1–2 per cent aqueous propylene phenoxetol solution (p. 124) or a few crystals of magnesium sulphate or menthol can be added. The animals should be left in the anaesthetizing solution until they cease to respond to probing, but they should not be allowed to die in the anaesthetic as tissue breakdown by autolysis can set in very quickly. When the animals are adequately anaesthetized the solution can be poured off and replaced by the fixative fluid. An alternative method of obtaining expanded holothurian material is to allow the animal to expand in a polythene tube of sea water and to kill it quickly by immersing the whole in hot water.

Echinoderms are best fixed in a 10–12 per cent solution of buffered formalin made up in sea water. Unbuffered formalin should never be used as free acids will quickly dissolve the calcareous skeletal elements. Material can also be fixed in 95–100 per cent alcohol, but this sometimes causes excessive shrinkage of the soft parts. Some species of feather-stars can only be prevented from disintegrating by plunging into strong alcohol. With large sea-urchins one or two small holes should be made in the side of the test or in the skin around the mouth to allow the fixative and preservative fluids to penetrate. Echinoderm material is usually stored in 70–90 per cent alcohol.

Hard-bodied echinoderms, for example, sea-urchins, brittle-stars and certain starfishes, can also be preserved by drying. Before drying, however, it is generally an advantage to fix material, since unfixed specimens tend to shrink and distort. Material for drying can be fixed in formalin as described above, but better results, for echinoids at least, are usually obtained with a corrosive sublimate (p. 131). The material should be placed in a non-metallic container and soaked for about 12 hours in a 3–4 per cent solution. After soaking, specimens should be washed repeatedly and then dried, preferably by being placed on a grid or mesh so that air can circulate around them.

The following method can be employed for the preparation of clean echinoid tests for examination of the detailed structure. The test is partially immersed (about two-thirds) in a commercial bleach solution such as Parozone or Domestos, or laboratory sodium hypochlorite, used neat or diluted according to strength (fresh stock may be very active but soon loses its strength after opening). It is advisable to do a test spot with undiluted bleach to determine its strength. If the entire shell or test of the urchin is required then the jaw apparatus and internal organs should be removed first by cutting round the soft peristome and scooping them out. Removal of the spines by scraping, scrubbing or pulling after an initial soaking in the bleach solution speeds the cleaning as the muscles at the bases of the spines take the longest to dissolve away. Also, the apical system is liable to drop away before the main part of the test is completely clean. As the apical system is so vulnerable it is advisable to only partly immerse the test in the bleach and to rely on capillary action to clean the skin from the apex. Caution is needed if the bleach is strong, because if the process is not checked the whole test is liable to disintegrate. Also, undiluted bleach dissolves human skin as well as that of sea-urchins so both must be washed thoroughly with plain water. The use of bleach to clean echinoid tests does not remove any natural colour patterns that they may have.

## Hemichordata (Plate 12b, e)

The Hemichordates are solitary or colonial, more or less worm-like invertebrates in which the body and its cavity (coelom) are divided into three regions – an anterior proboscis, a collar and an elongated trunk. They are exclusively marine and the phylum is made up of three classes – Enteropneusta, Pterobranchia and Planctosphaerida.

The Enteropneusta or acorn-worms are relatively large solitary animals usually 100–150 mm long (but rarely 1 or 2 m) with a blunt, often acorn-shaped, proboscis. Behind the collar, the trunk bears on each side a longitudinal row of gill slits. All are shallow-water species. Some are found under stones and amongst seaweeds but a number of species

Plate 12 Chaetognatha, Hemichordata, Chordata
a, *Branchiostoma*, an amphioxus or lancelet; b, *Saccoglossus*, an acorn-worm; c, *Ciona*, a solitary sea-squirt; d, *Sagitta*, a planktonic chaetognath, or arrow-worm; e, *Rhabdopleura*, part of a colony, 5 mm; f, *Botryllus*, a colonial sea-squirt.

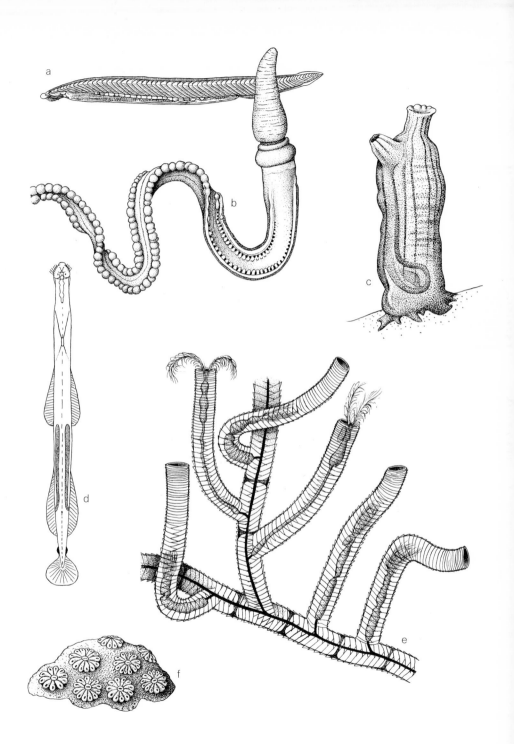

85

live concealed in burrows in mud or sand. The enteropneusts are ciliary feeders, collecting food particles outside the alimentary tract through the agency of the cilia of the body surface and its mucous secretion. Some species emit a characteristic odour reminiscent of iodoform. The geographical distribution of enteropneusts is imperfectly known, but they appear to favour warm and temperate waters and several species are known from European coasts.

The Pterobranchia are hemichordates in which the collar region carries two or more tentacle-bearing arms. Gill slits may or may not be present. The majority are colonial forms that live in secreted tubes (coenicia) attached to the bottom in relatively deep waters. Most records of Pterobranchia are from the southern hemisphere although some are known also from northern latitudes notably *Rhabdopleura*, the thread-like tubes of which are most often found encrusted on dead shells.

The Planctosphaerida are known only from planktonic larvae.

### Collecting
Enteropneusts can be obtained by the usual methods of shore collecting (p. 96), and on mud and sand their presence is often betrayed by spiral castings. Pterobranchia are usually obtained by dredging, and *Rhabdopleura* can be found on shells and rocks with Bryozoa and sea-squirts.

### Preserving
It is generally an advantage to anaesthetize hemichordates before fixing them, and this can be done by the gradual addition of 95 per cent alcohol to the sea water containing them until a concentration of 1 part of 95 per cent alcohol to 9 parts of sea water has been attained. Enteropneusts can also be anaesthetized by immersing them in a 7 per cent aqueous solution of magnesium chloride.

Hemichordates can be fixed in a 5 per cent sea water/formalin solution, but if material is required for histological study sea water/Bouin's fluid (p. 128) is to be preferred. After fixation for about 24 hours, formalin-fixed material should be transferred to fresh 5 per cent sea water/formalin for storage. Bouin-fixed material should be stored in 70 per cent alcohol.

## Chordata (Plate 12a, c, f)

This phylum is composed almost entirely of vertebrate animals. However, the members of two sub-phyla (Urochordata and Cephalochordata) are invertebrate although they possess at some point in their life cycle three chordate attributes – a notochord, a dorsal tubular nervous system and pharyngeal slits.

## Urochordata (Plate 12c, f)

As adults the Urochordata (known also as tunicates) bear little resemblance to other chordates. The majority are sessile barrel-shaped animals and the chordate features are distinct only in their free-swimming larvae which resemble minute tadpoles. They are exclusively marine animals and are placed in three classes – Ascidiacea, Thaliacea and Larvacea.

The Ascidiacea or sea-squirts are sessile tunicates that may be found as solitary forms (single sac-like animals with two prominent siphons) or as colonial forms attached to, or encrusting, rocks, stones or seaweeds. The common name derives from their habit of squirting water and excretory matter through the exhalent siphon. Ascidians vary considerably in colour, grey and green species being quite common. In the solitary forms the tests can be smooth or quite mammilated, and in consistency they can be soft, firm or leathery. The configuration of the colonial forms also varies markedly. In the simple colonial species more or less discrete individuals are joined by stolons, but in the case of the more specialized colonial species the individuals comprising the colony are embedded in a common test. Both colonial and solitary forms may be encrusted with debris (shell fragments, etc.) and by other sessile organisms.

Ascidians are cosmopolitan in distribution. Many species can be found in the intertidal zones, but the majority are subtidal and have even been found at great depths.

The Thaliacea and Larvacea are more-or-less transparent planktonic animals. Thaliacea (salps, doliolids and pyrosomids) occur usually in shoals, principally in tropical and sub-tropical waters (some species are colonial and most are brilliantly luminescent) while the small tadpole-like Larvacea are found in plankton throughout the world.

### Collecting
Solitary ascidians can usually withstand quite a lot of rough handling and can generally be prised off the substratum with a blunt knife. Wherever possible colonial forms should be collected still attached to their immediate substratum, and a hammer and chisel may be necessary to chip off portions of rock carrying encrusting species.

**Preserving**

Ascidians should be anaesthetized before being killed and fixed. The animals should be placed in a bowl of sea water and a few drops of a 1–2 per cent solution of propylene phenoxetol or a few crystals of magnesium sulphate or menthol added. After anaesthetization they can be killed and fixed by adding buffered commercial formalin solution to the water containing them. As an alternative to chemical anaesthetization, ascidians can be left in a bowl of sea water for several hours until they are fully expanded and partially asphyxiated. As before, they can then be killed by the addition of buffered formalin solution to the water. Sessile tunicates are best preserved in 70–90 per cent alcohol but buffered 5 per cent formalin can be used. Pelagic tunicates can be killed, fixed and preserved by plunging them into 5 per cent formalin but better results are obtained if they are first anaesthetized and then fixed in Schaudinn's (p. 132) or Bouin's fluid (p. 128). Material fixed in Schaudinn's fluid should be thoroughly washed before being preserved in 5 per cent formalin or 79–90 per cent alcohol.

# Cephalochordata (Plate 12a)

Cephalochordates are small laterally compressed animals that occur in shallow marine waters all over the world. One of the best-known species is the lancelet, *Branchiostoma* (formerly *Amphioxus*) *lanceolata*, which occurs in European waters. Cephalochordates can swim freely by fish-like undulations of the body, but they are mainly found partially buried in mud or sand with only their anterior ends protruding. In this position they feed by drawing water into the mouth and filtering off suspended micro-organisms.

The Amoy amphioxus, *Branchiostoma belcheri*, which is particularly abundant along the south-eastern coast of China, is fished commercially and about 35 tonnes are landed each year. The common lancelet, *B. lanceolata*, is found in shallow seas in many parts of the world and in British waters it occurs in shelly bottoms near the Eddystone lighthouse and on the Dogger Bank in the North Sea. Immature stages of amphioxus are commonly found in plankton hauls, but there is evidence to suggest that larvae descend to deeper water during the day, and that they live mainly on the sea bottom. A type of giant, pelagic, larval amphioxus known as the Amphioxides larva, in which the gonads develop before metamorphosis, has a wide distribution.

## Collecting

Adult cephalochordates can be taken with simple dredging equipment and the larvae by tow-netting. One particular dredge, which has a rigid circular mouth-opening particularly suitable for collecting on shell and gravel deposits, is called the 'Amphioxus dredge'.

The larvae show a marked diurnal migration and can be collected just after sunset in the upper layers of the sea.

## Preserving

For general studies cephalochordates should be fixed and preserved in 5 per cent buffered formalin.

Chapter 2

# Collecting methods and apparatus

Many invertebrates can be collected by simple direct means. Thus, larger animals can often be captured by hand or with a pair of suitably sized forceps, and many small species can be easily collected in adequate numbers with a moistened camel-hair brush. However, for some animal groups in particular biotopes, special methods and equipment are needed. The techniques are generally not difficult and the apparatus can often be improvised or constructed with inexpensive and easily obtainable materials. Some of the more important special techniques are described below under general headings referring to types of habitat.

## General requirements

*Bottles and tubes*   Specimen tubes of various sizes with stoppers will be required. Plastic tubes, being light and unbreakable, are generally more convenient and in some sizes are available with hinged or self-sealing stoppers. Wide-mouthed screw-capped jars will be needed for larger specimens. Very large specimens are best kept in tin-plate or plastic containers with tight-fitting covers, but tin-plate vessels, which are usually sealed by soldering, should not be used for specimens preserved in formalin solution since this preservative tends to corrode metal.

*Collecting instruments*   Generally useful are forceps of various sizes, camel-hair brushes, pipettes with rubber teats, dissecting scissors and knives. Wooden forceps are often useful as they do not rust, and they can be made to almost any size by binding two flexible laths of wood to a small block with copper wire.

*Dishes*   Shallow glass, porcelain or plastic dishes are useful for containing fluids in which animals are to be sorted, washed or killed, but ordinary plates or saucers will do equally well. The white trays used in photographic darkrooms are ideal for this purpose. Enamelled metal vessels are particularly useful when travelling and for use with hot liquids.

*Tools*   A small trowel is often worth carrying and a cold chisel, hammer, strong penkife or similar instrument can be handy for prising off bark, loose rock fragments, etc.

*Lenses*   A good hand-lens is indispensable, and aplanatic lenses are the best. The most generally useful magnifications are $\times 10$ or $\times 15$, but it is often convenient to have a pair of lenses of, say $\times 10$ and $\times 20$, mounted in the one holder. As lenses are easily mislaid, or laid down and forgotten in moments of excitement or despair, they should always be carried on a lanyard, attached to the person.

*Miscellaneous*   A good supply of plastic bags of various sizes should be taken on any collecting trip as their uses are unlimited. Other important items are elastic bands, string, thread, gum, cotton wool, muslin, tin boxes and pill boxes. Rubber gloves are often useful, particularly for handling specimens in formalin or for clearing away 'hostile' vegetation. One thing which must never be forgotten in an adequate supply of suitable labels and the necessary writing equipment – few things are more useless than a collection of animals from an unknown locality. Finally, a haversack of some kind will be required to carry these materials, and it is as well to remember that the return load is likely to be a great deal bulkier than the one set out with.

*Photography*   Good colour photographs of the living animals can be extremely useful as colour patterns are often a valuable aid to identification and are frequently lost during fixation and preservation. Also, colour photographs of the immediate habitat from which a collection is taken may provide important information.

## Free-living animals

### (a) Terrestrial habitats
*Entomological equipment*   Apart from the insects the most common invertebrates living freely on bare ground and amongst vegetation are the arachnids (for example spiders, harvestmen and mites), and in these biotopes free-living animals are generally collected by using entomological techniques. Animals in grass and low herbage can be taken with a sweeping-net and the denizens of trees and shrubs can be examined by using a beating-tray which is laid on the ground while the branches directly above are knocked sharply with a heavy stick. Entomological aspirators, or 'pooters' (fig. 1) are simple but effective devices for

sucking up specimens from nets and trays and they can often be used for direct capture. Sweeping-nets and beating-trays are obtainable from dealers in entomological requisites and further details of these methods are given in booklet No. 4a of the series, *Instructions for Collectors*, published by the British Museum (Natural History).

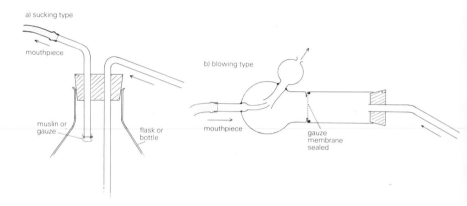

a) sucking type

mouthpiece

b) blowing type

muslin or gauze

flask or bottle

mouthpiece

gauze membrane sealed

Fig. 1 Entomological aspirators or 'pooters'
a, sucking type   b, blowing type
*Note*  In most cases the sucking type of aspirator can be used safely, but in circumstances which might present a health hazard the direct-sucking method must be avoided. As an alternative a rubber suction-bulb can be attached to the pooter's mouthpiece, or a blowing type of aspirator (Fig. 1b) used instead.

*Desiccation funnels*  Equipment in this broad category is used for the automatic collection of animals (mainly Arthropods) living freely in soil, litter, dung and in organic debris of all kinds. Basically these devices are very simple. A sample of the material from which the animals are to be extracted is spread fairly thinly on a sieve held over the wide opening of a funnel (fig. 2). As the material dries out the animals move downwards until they eventually drop through the funnel into a collecting tube held below the lower opening. The method was invented by an eminent Italian zoologist, Antonio Berlese – hence Berlese funnel – and as originally devised the apparatus consisted of a large metal funnel equipped with a hot-water jacket, the heat from which served to dry out the material on the sieve.

Modifications of the original apparatus are seemingly endless. In the

Tullgren funnel, for example, an electric-light bulb replaces the water jacket as a source of heat, and it is alleged that the light acts as an additional stimulus to downward movement since most of the animals are negatively phototropic. Other devices in this broad category – steep-sided funnels, high-gradient funnels, split funnels – incorporate modifications important for quantitative work, and a detailed account of this sort of equipment is given by Murphy (1962). For general qualitative

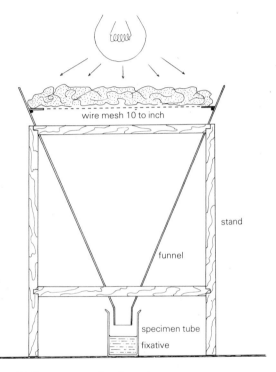

wire mesh 10 to inch

stand

funnel

specimen tube

fixative

Fig. 2 A Berlese–Tullgren extraction funnel

collecting in the field, mass-produced plastic funnels 215 mm in diameter are quite adequate. The material under examination is held on a circle of metal gauze 165 mm in diameter with a regular square mesh of 3 mm and dried by suspending a kerosene lamp above the funnel. In hot dry climates an artificial heat source may not be necessary and even in temperate regions very useful collections can be obtained by allowing the material to dry out 'naturally' over a period of several days. The progress of the extraction can be checked by inspecting the contents of the collection tubes. A number of Museum expeditions have dealt

Plate 13 Desiccation funnels set up in bamboo frame.

with soil and litter samples by mounting a series of plastic funnels on frames constructed with locally obtained materials. Since the samples for processing generally store quite well for several days – or even weeks – in tightly closed polythene bags, it is often convenient to set up funnel assemblies at a base camp from which collecting sorties can be made. On one British Museum Himalayan expedition, for example, soil, litter and dung samples were processed on 53 funnels set up on a bamboo frame in a rough hut at a central base camp (Plate 13). Heat for drying was obtained by suspending kerosene pressure lamps above the assembly, and the material was left on the funnels for at least 10 days.

*The Baermann funnel*  This technique is used mainly for isolating free-living nematodes from soil and organic debris, and plant parasitic nematodes from small pieces of infested tissue, but it has also been used for collecting Rotifera, Enchytraeidae, Turbellaria and Tardigrada; although in the case of the latter the value of the technique is doubtful. The apparatus consists of a glass funnel about 100 mm in diameter in which the material, suitably weighted if necessary and contained in a piece of butter muslin, is placed on a metal sieve resting in the funnel. A short piece of rubber tubing, closed by a spring-clip, is attached to the funnel stem, and warm water (40–42°C) is added to the funnel until it comes into contact with the sample. The warmth stimulates the animals so that they move out of the sample into the water. Within a few hours they sink to the bottom of the funnel stem where they can be run off. Various modifications of this technique are described by Murphy (1962).

*Flotation*  The active stages of soil arthropods are generally quite easily collected by heat-desiccation techniques, but these methods do not work too well with the 'heavier' mineral soils, that is to say mineral soils containing a high proportion of small particles belonging to the clay fraction. On drying, heavy loams and clayey soils tend to set into a hard mass so that animals living in them are unable to escape. The fauna of such soils can be extracted by flotation methods, but these techniques are elaborate and rather tedious, and generally quite unsuitable for expedition work. However, if some basic laboratory facilities are available the extra effort could be worthwhile since the communities of heavy mineral soils have been little studied and material collected is likely to be especially valuable.

Essentially the flotation method consists of stirring the soil in a magnesium sulphate solution of specific gravity 1·1 and passing a stream of air bubbles through the suspension from below. The air bubbles keep the soil particles in suspension, thus freeing animals and other light material so that they can rise to the surface of the liquid. Under field conditions air can be bubbled in from a car foot-pump, or better still from an air cylinder if one is available. After stirring and 'bubbling', the soil suspension is allowed to settle, and the float, which consists of animals and organic debris, is decanted onto a 100- or 120-mesh sieve and washed gently with water. The animals in the float can be separated from the plant debris by an oil-flotation technique, but this requires fairly elaborate laboratory facilities and the best that can generally be done under field conditions is to wash the organic debris/animal mixture with alcohol from the sieve into a collecting tube.

There are several variants of the basic magnesium sulphate flotation method and a modification designed specifically for collecting micro-arthropods is described by Raw (1955).

**(b) Marine habitats**
The problems involved in the collection of marine organisms are many, and a detailed treatment of the subject would more than fill this small book. Only on the sea-shore is collecting a relatively simple task, requiring little specialized equipment, at least for the larger animals. The smaller organisms living in the interstices of the substratum and constituting the meiofauna and microfauna can be extracted without too much difficulty using techniques similar in principle to those employed in the study of terrestrial soil and litter fauna. In most cases the apparatus is very simple and can often be improvised in the field. It may well be necessary to adapt individual methods to suit particular working conditions especially in relation to animal size and sediment type. Those techniques that are most commonly used for work on the shore are described below under suitable headings.

An enormous variety of tools has been used for collecting animals in the sea or from the sea-bed. Since man himself cannot penetrate the aquatic environment beyond the limits of free-diving, it is necessary to collect information and samples by 'fishing' from the surface. The rapid expansion of oceanographic and fisheries research over recent years has led to the development of a great array of 'fishing tackle' with seemingly endless modifications to suit particular research programmes.

Many of these devices are far too complex and expensive for the general collector and the present account is confined to the description of relatively simple techniques which might reasonably be used for example on general expeditions and college field courses.

## (i) Intertidal macrofauna

The sea-shore offers the most readily accessible part of the marine environment as in most parts of the world it is regularly exposed during the tidal cycle. If possible, collections should be made at the time of the spring tides which occur every fortnight and which relate to the arrival of the new and full moons. It is during these tides that the sea level falls to its lowest point, thus exposing the extreme low levels of the shore that remain covered through the intervening neap tides.

The methods employed for collecting in the intertidal region are essentially simple and depend largely on the type of shore to be studied. It is important to remember that many of the animals on the shore move into hiding when the tide recedes, especially during the daylight hours, and therefore all hand-collecting must be carried out thoroughly.

On open rocks, non-sessile animals that are large enough to be seen with the naked eye or with the aid of a hand-lens can be picked off by hand, with blunt forceps, or with a fine paint brush. On the other hand, many of the sessile animals found on the rocks would be badly damaged if pulled indiscriminately from the surface to which they adhere. If possible, therefore, small pieces of rock should be broken off with the animal still attached. In circumstances where this is not feasible a thin-bladed knife should be slid between the organism and the rock at its point of attachment and the animal carefully levered away from the surface.

A number of motile animals retire under a protective covering of seaweed when the water recedes, and these need to be looked for by carefully displacing the weed. Since the seaweed itself provides a refuge for many motile and sessile creatures, a variety of weeds should be removed from the rock for examination. Those parts of the weed that carry colonial animals should be retained separately, while the remainder is washed thoroughly in a bucket of sea water. This treatment will cause many of the motile animals to be flushed from their hiding places, and if time permits the material should be allowed to stand for a few hours until the water becomes stale. Under these conditions some of the animals not dislodged by washing will detach from the weed and collect on the sides and bottom of the container. They may be further induced to detach from the weed by the addition of some fresh water, a small quantity of formalin solution or alcohol, or by the addition of a few crystals of magnesium sulphate (Epsom salts). Finally, any compacted growths such as holdfasts should be carefully teased apart to release any remaining animals.

Apart from hiding in weeds some animals will retire into rock crevices during the low tide. These animals can sometimes be induced to leave the crevice by squirting in a weak solution of formalin, but the sessile fauna can only be reached by breaking open the crevice with a crowbar or with a hammer and cold chisel.

For collecting from relatively flat surfaces, such as wharf piles and steep rock faces, a long-handled scrape-net is very useful. This is usually a D-shaped net with a scraping edge along the flat side.

Rock pools should receive careful attention as they often contain a rich fauna and flora. Most animals found in rock pools can be removed by the methods already described but a fine-mesh hand-net and a wide-bore pipette are useful for capturing fast-moving species.

Several groups of animals are associated with gravels, sands and muds. By turning over large stones and boulders lying on loose sediments a surprising variety of forms may be revealed, including those living in burrows. Very few species, however, are normally visible on the surface of soft sediments and must be extracted from it. This is most easily accomplished by digging with a fork or a narrow spade, sifting the sediment through the fingers to remove the larger organisms, and then through sieves to retain the smaller ones. Sieving for the microscopic animals may often be accomplished more effectively in the laboratory if it is convenient to bring back the sediment sample. It is often desirable to make quantitative estimates of the fauna and for these a sediment sample of known area and depth must be taken.

Often, after sieving, large quantities of inorganic particles remain with the animals. Hand-sorting of animals from this debris is very time-consuming and use can be made of a flotation technique described by Birkett (1957). The sample is first stirred thoroughly in a liquid such as carbon tetrachloride, and then allowed to settle. The mineral matter in the sample will sink to the bottom but as the specific gravity of the animals is less than that of the solution they rise to the surface and can be removed easily. This method is only suitable for extracting animals from sediments such as shell, gravel and sand which contain little organic debris. In organically rich sediments, such as mud, the debris will float to the surface along with the animals. In these circumstances the specific gravity of the fluid has to be controlled more closely in order to separate various organic fractions eventually floating off the animals. Anderson (1959) found that a solution of granulated sugar in water adjusted to a specific gravity of 1·12 (approximately 1 kg granulated sugar per 4·5 litres of water) allowed most of the detritus to sink.

Methods of collecting small animals and the young stages of larger animals that live in the interstitial spaces in the surface few centimetres of soft substrata in the intertidal region are described in the next section.

## (ii) Meiobenthic fauna

This element of the benthic infauna comprises those small organisms that live in the interstitial spaces of a soft sediment. A precise definition of the group is difficult, and probably unnecessary, since they form only a size class between the macrofauna and the microfauna. They can be suitably described as organisms small enough to move through the interstices of a sediment without disturbing the sediment particles, or on a purely size basis as that part of the infauna that will pass through a 1 mm-pore filter. The study of the meiofauna has attracted considerable attention in the last few years with the consequent development of certain specialized collection methods. A very full account of the procedures and techniques is provided by Hulings & Gray (1971) in a manual on meiobenthology. This manual was compiled following the International Conference on Meiofauna held in Tunisia in 1969, and covers many aspects of research in this field for the benefit of both the specialist and the non-specialist. The authors recommend the adoption of certain standard quantitative methods with the hope that these will be followed by all research workers making it possible to draw direct comparisons between the results. All too often it is impossible to compare results directly because different procedures have been adopted in the collection of data.

When the tide is out the simplest method of collecting interstitial animals is to dig a small pit in the sediment, with a number of small channels radiating over a distance of 2–3 metres, and to allow water to accumulate in it. It may assist the drainage if the surrounding area is disturbed with a fork. To filter off the free-swimming animals water from the pit is then poured through a fine sieve (62 $\mu$m-pore diameter) and a sample of the sediment from the bottom of the pit is examined under a binocular microscope for the less active forms. This technique will provide some insight into the meiofauna of a particular area, but may not liberate all groups of animals present, especially those which adhere strongly to the sediment particles. Furthermore, most investigations require a quantitative approach to the subject and necessitate a more effective method for extracting the animals.

Since the meiofauna is numerically very rich, only a small sample of sediment is needed and this is best obtained with a simple corer, which can be improvised from a length of plastic or brass tubing about 400 mm in length and 30–40 mm in diameter. The tube is pushed or hammered the desired distance into the sediment. The upper end is then sealed with a rubber bung and the corer withdrawn. By easing the bung gently to allow air to enter the tube the core can be made to slide out. Apart from the simplicity of operation, coring provides a means of obtaining a standard volume for quantitative studies, and the core can be readily sub-divided into a number of depth zones for investigating the vertical distribution or vertical migrations of the organisms. A number of methods of extracting the meiofauna from the sediment sample are given below.

*Filtration* A quantity of sea water is filtered through a 62 μm mesh into a bucket and stirred up thoroughly with the sediment sample. After allowing the suspension to settle momentarily, the water is decanted through a 62 μm net, and the procedure repeated a number of times. Finally, the filtrate that contains the animals and some of the lighter organic debris is washed into the cone of the net, and the net everted into a suitable container. The animals can be preserved in 10 per cent formalin solution or 70 per cent alcohol, but in each case it should be added drop-wise over a period of several minutes. The process of simple filtration offers a reasonably effective method of extraction and is particularly recommended for sandy sediments.

Both this method and the bubbling method mentioned below can be made more effective by the addition of an anaesthetic, such as magnesium chloride, alcohol, formalin solution or propylene phenoxetol, to the filtered sea water. When the sediment sample has been stirred in, it must be left for some 10–20 minutes to allow time for the relaxant to take effect.

*Bubbling method* This flotation technique was first used by Higgins (1964) and is especially suitable for fine sediments such as mud. The sample is mixed thoroughly with filtered sea water containing a suitable anaesthetic, and the suspension agitated further by bubbling air into the mixture. Under field conditions the air can be supplied from a bicycle pump or a car foot-pump. As the bubbling continues, a 'float', containing animals liberated from the sediment together with some organic debris, collects on the surface of the sea water. The water is then allowed to stand for a moment before the surface is blotted with a piece of paper to remove the 'float' and the paper washed into a 62 μm net. The extraction needs to be repeated several times before the contents of the net are washed into a collecting dish and preserved in the manner described above.

*Seawater-ice technique* This method was first described by Uhlig (1964) and provides effective extraction of meiofauna from porous sediments such as sand. A sample of sediment is placed at the bottom of a tube on a fine nylon mesh (140 μm-pore diameter) and covered with a thin layer of cotton wool (fig. 3). Over this, the tube is filled with crushed seawater ice. As the ice slowly melts it drains through the sediment sample extracting the small organisms into a collecting dish. If ice is not available it is possible to achieve satisfactory results using cold water, so long as a fine filter can be placed over the sample to allow the water to percolate slowly. On the other hand, water at 40°C

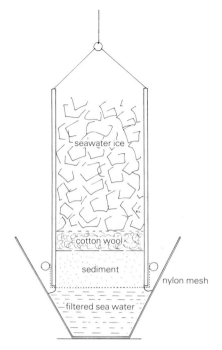

Fig. 3 Uhlig seawater-ice technique (after Uhlig, 1964)

has been used with some success and this is probably a useful alternative when working in tropical climates.

*Boisseau elutriation methods*    This is a laboratory technique (Boisseau, 1957) eminently suitable for preserved samples but less effective with some living organisms which may adhere to the walls of the apparatus. The Boisseau apparatus is shown in fig. 4. The sediment is placed in the separating funnel together with a suitable anaesthetic, and the latter given time to take effect. The sample is then stirred up by passing a stream of filtered sea water through the tap on the separating funnel, and the flow of water adjusted to give adequate agitation without flushing the sediment out of the glass column. Animals washed out of the sediment are carried by the water into the descending limb of the apparatus from which they are removed from time to time by opening the tap, and filtered over a 62 $\mu$m mesh.

Certain groups of animals recover very quickly from the effects of the anaesthetic as soon as the seawater circulation is started and attach themselves firmly to the sediment particles or the walls of the apparatus. This can be prevented by adding the anaesthetic to the circulating sea

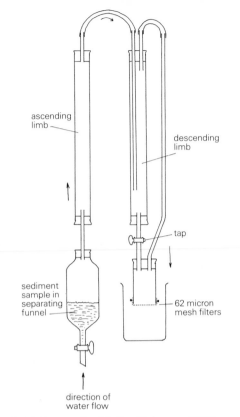

ascending
limb

descending
limb

tap

sediment
sample in
separating
funnel

62 micron
mesh filters

direction of
water flow

Fig. 4 Boisseau elutriation method (after Boisseau, 1957)

water as well as to the sediment sample, so that an adequate concentration is maintained throughout the elutriation process.

## (iii) Sub-tidal benthic fauna

In this section some of the apparatus and techniques available for collecting animals from the sea-bed are described. It will necessarily be restricted to relatively inexpensive methods using readily available apparatus since the great variety of sophisticated and costly gear used in oceanographic and fisheries research is far outside the scope of this book.

In calm weather the animals of the very shallow sub-tidal region can be reached by wading into the water, and collections made by hand using the methods recommended for the shore. If it is difficult to see into the water, an open glass-bottomed box placed against the water surface will eliminate surface ripples and give a clear view of the bottom.

### Diving

In depths less than 50 metres the diver equipped with self-contained underwater breathing apparatus (SCUBA) can make excellent collections of marine invertebrates. Free-diving is a particularly effective method of collecting from areas of hard substrata where the more traditional methods of 'fishing' using trawls, grabs, etc., are unsatisfactory. The diver can probe directly into crevices, caves and canyons, examine steep rock faces and overhangs, turn over boulders, and study animal communities found under the seaweed and in the weed holdfast.

Although it is possible to work at depths of 50 metres using the conventional aqualung, the time available at this depth is very limited and permits only brief periods of work. A maximum depth for a reasonable period of collecting would be about 30 metres and for long periods of work only 15 metres. Beyond this, extra air would be required to allow time for decompression. The air supply does, in fact, impose the main restriction on the length of a dive, and the more work the diver does the more air he will use.

On open rock surfaces animals can be picked off with blunt forceps or, if sessile, detached with the aid of a thin-bladed knife, and placed in polythene containers filled with sea water at the beginning of the dive. A small-mesh nylon shopping bag makes an ideal underwater haversack for carrying the collecting bottles. For making notes under water a matt-surfaced plastic note-pad with a soft pencil attached is quite satisfactory.

Active invertebrates may be lured into capture by placing suitable bait, such as crushed sea-urchin or crushed mollusc, inside a container on the sea floor; when the animals have entered the container to feed the lid is replaced. Trapping methods can also be used from the surface. The trap, which has some mechanism to prevent the escape of animals once inside, is lowered to the bottom together with a bait such as fish offal. An ordinary lobster pot is a good example of such a trap. After a suitable period the pot, which has been marked with a buoy, is hauled up and the catch withdrawn.

Errant animals hiding in narrow crevices can sometimes be driven out by directing a stream of commercial formalin solution, alcohol (70 per cent) or an anaesthetic at them from a plastic squeeze bottle. A tool known as a 'podger' or crab-hook is handy for extracting large invertebrates such as crabs, crawfish and lobsters from crevices. Whilst distracting the animal's attention the podger is slid underneath the body

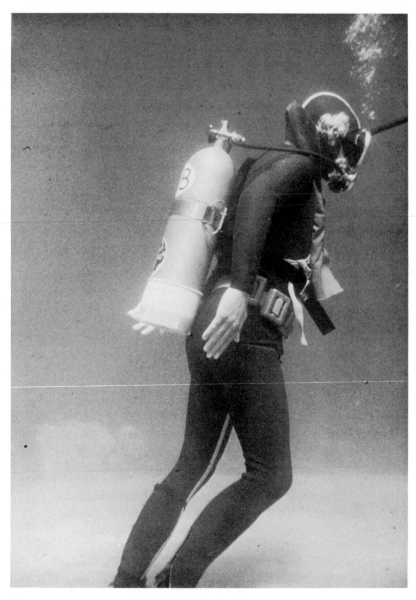

Plate 14  SCUBA diver on sea bed

and lodged either on the ventral side or around the rear of the carapace. The animal is then pulled slowly out of its hole. Another instrument with a curious name, the 'slurp-gun' is useful for catching small active animals without damaging them. The gun (fig. 5) is made from a large-bore plastic tube with an inner piston attached to a handle, and a smaller bore tube to act as a barrel. Using the gun in the manner of a syringe, small animals can be sucked up the barrel and then blown into a polythene bag or other suitable container.

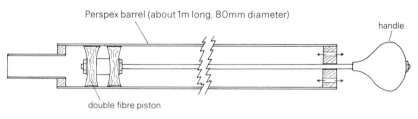

Fig. 5 A slurp-gun

On soft sediments a diver is able to pick off the dermersal fauna with great precision, but in the case of very fine sediments such as mud care must be taken not to stir up the bottom as this will soon reduce visibility. Animals living in the sediment can be collected with a diver-operated air-lift sampler. This apparatus consists of a rigid wide-bore tube which has a high-pressure air pipe entering the tube near its mouth. Compressed air from an attached cylinder, or from a surface supply, is introduced into the tube and the stream of bubbles rising up the tube produces a suction effect below. In this way soft sediment and resident animals can be sucked up the tube and passed on to a sieve at the outlet from which the filtered animals then fall into a sample bag. A design for a portable suction sampler is given by Hiscock & Hoare (1973).

*Dredges and trawls*   These tools are used mainly to obtain qualitative samples of epifauna, but because of their limited penetration into the sediment they do not provide much information about the infauna. The device most frequently used for making collections of the epifauna, and one which will operate over most bottom conditions, is the Naturalist's or Rectangular Dredge. This device consists basically of a strong rectangular metal frame, which forms the mouth of the dredge, to which is attached a small-mesh net (fig. 6). The dredge has two hinged towing arms which are tied together at the point of attachment for the tow-rope. Only one arm is fixed directly to the tow-rope, the other is bound to it with twine to form a weak link in case the dredge becomes fast in the sea-bed. If this happens, hopefully the twine will break under the extra strain thus allowing the arms to separate and the dredge to be freed sideways. This type of dredge is quite easy to handle and it does not matter which side is up when it reaches the bottom. It should be towed very slowly (1 knot), indeed if there is sufficient current it may just be possible to let the boat drift for 5–10 minutes before hauling in. On very soft sediments, such as mud or loose gravel, the dredge will fill up very quickly and can be hauled in almost at once.

Naturalist's rectangular dredges, which come in a variety of sizes from 30 to 75 cm wide, can be obtained from the Marine Biological Association Laboratory, Plymouth. The small 30 cm frame dredges are suitable for hand-hauling from a small boat, but on boats with power winches the larger sizes can be used. The gauge of mesh forming the bag of the dredge can be varied to suit the type of sample required.

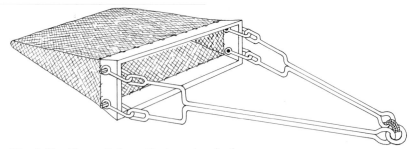

Fig. 6 The Naturalist's or Rectangular dredge

If a sample of the sediment is needed, an inner bag made from stramin can be incorporated. Again, if the bottom is known to be particularly rough and rocky the dredge bag can be made of metal rings or wire as in the familiar oyster and scallop dredges. Round or curved frames have been employed to achieve some penetration into the sediment to collect the infauna, but their effectiveness is limited, except perhaps on soft mud, and the infauna can be more effectively sampled with a grab or with an Anchor Dredge.

As the name implies the anchor dredge is specially designed to dig itself into the bottom substratum, and it is a most effective tool for sampling sands and similar finely packed sediments. The dredge, first described by Forster (1953), consists of an open metal box, one side of which is inclined as a cutting plate. A pair of stout arms is attached to the upper side of the dredge and a strong net forms the collecting bag. The back end of the bag is open but tied securely with rope during the operation so as to facilitate emptying the net when it is hauled

Plate 15 Anchor dredge about to be emptied on-board ship

on board. The dredge is shot with the boat moving slowly ahead or astern, usually just drifting, and as the strain is taken on the tow-rope the dredge digs into the bottom as would an anchor. To release the gear from the sediment the boat steams slowly in the opposite direction while the warp is hauled in so that the dredge is broken clear of the bottom as the boat passes above it. Clearly, this equipment can only be operated from a vessel equipped with a power winch. Holme (1961) describes a modification of the Forster dredge referred to as a double-sided anchor dredge. This has the advantage that it will operate either

Plate 16 Agassiz trawl on deck

way up – the towing arms swivel at the base depending on which way the gear hits the bottom. This is a particularly useful aid when working over deep water where the attitude of the dredge is difficult to control. The front edge of the dredge has a symmetrical metal opening without the sloping cutting edge of the Forster dredge. Consequently it will not bite into the bottom quite as cleanly and has to be pulled along somewhat to drag up the sediment.

Many epibenthic invertebrates, particularly the larger, scarcer, or faster-moving species can be taken by trawling. Unlike a dredge, which scrapes or digs at the sediment, the trawl is designed to skim over the

108

sea-bed catching the animals on or close to the bottom. Because of their large size the familiar 'otter' and 'beam' trawls used by the commercial fisheries are not usually effective for capturing small invertebrates, and for general scientific collecting a much more useful device is the Agassiz trawl. This trawl consists simply of a light metal frame with rounded runners at each end, attached to a small-mesh collecting bag. A pair of rope bridles is used to join the frame to the towing rope, and the trawl is pulled along at speeds of about 3 knots for a period of some 10–15 minutes. Agassiz trawls can be obtained from the Marine Biological Association, Plymouth, in frame sizes from about 1250 mm up to 2000 mm.

*Grabs and corers*    Apart from the deep digging anchor dredges that can 'bite' into the bottom some 200 mm, and which will sample the deep-burrowing infauna, most dredges explore only the surface few centimetres. If quantitative samples of the sediment are required, grabs are generally employed in preference to dredges. They do, of course, catch only the sedentary or slow-moving animals, since the more active species are able to escape the slowly closing jaws. Like dredges, grabs are available in a wide variety of forms, but all operate on a common principle; they are lowered vertically from a stationary boat to rest on the sea-bed from which they take a bite of known size and shape. Naturally, the penetration of the bite depends on the hardness of the deposit.

The Petersen grab is a simple apparatus which has been widely used for sampling soft sediments. It consists of a pair of heavy jaws that are locked apart whilst the grab is being lowered. The grab sinks into the bottom deposit under its own weight and as the cable falls slack the lock on the jaws is automatically released. On hauling, the jaws are drawn together by the cable, and a semi-spherical bite is taken out of the sediment. A worthwhile improvement of the Petersen grab is the van Veen grab which has a long arm attached to each jaw to increase considerably the leverage applied by the closing mechanism.

In addition to grabs which rely upon their own weight to penetrate the bottom deposit, a variety of spring-loaded samplers has been developed. One, which is generally recommended, is the Smith–McIntyre or Aberdeen grab (Smith & McIntyre 1954). This has a heavy metal frame housing a pair of spring-tensioned scoops. The apparatus is lowered to the bottom with the scoops open, and as the frame comes to rest a pair of trigger-plates is depressed to activate powerful springs which drive the scoops into the sediment. The scoop is finally closed as strain is applied to the hauling cable. The top of each scoop is covered by a wire mesh to prevent the sample from being washed out during recovery.

If quantitative collections of the meiofaunal element from offshore bottom sediments are required it is only necessary to take relatively small samples of the deposit. For this a device which will take an undisturbed core from the substratum to a depth of 150–200 mm may be more suitable than a dredge or grab. Such a device is the Moore & Neill core-sampler (Moore & Neill 1930). It is essentially a protected glass tube through which water flows freely during descent and which is driven into the mud upon impact. On hauling, a simple valve mechanism closes the top of the tube to prevent loss of the sample. The extraction of the interstitial organisms has already been dealt with in the section headed, meiobenthic fauna.

During recent years considerable advances have been made in the study of the offshore marine benthos, and a host of rather sophisticated sampling methods has been developed. The majority of these tend to lie outside the scope of the general collector but further details can be found in Holme (1964) and Holme & McIntyre (1971).

## (iv) Plankton

The term 'plankton' refers in general to those organisms living freely in the water column which, because of their limited powers of locomotion, drift with the water currents. This does not mean that these organisms are all floating passively; many swim quite vigorously but, because of their small size, the horizontal distance covered is relatively small. Also, much of their swimming effort is concerned with changes in vertical distribution as seen, for example, in the diurnal migrations.

Almost all plankton collecting methods depend on the same principle, that of separating the plants and animals from the water by filtration through some form of meshwork or netting. Usually the net is pulled through the water from a boat, but where a strong tidal current passes a landing stage or pier, a small plankton net can be fished quite effectively by allowing it to stream in the current.

The simplest and most widely used plankton collector is the tow-net (fig. 7). This consists of a cone of net, attached at the wider end to a canvas collar, which in turn is attached to a circular metal hoop. The hoop carries three rope bridles that are fixed to the towing warp some distance in front of the mouth of the net. The narrow end of the net cone also has a canvas collar that is tied around the neck of a glass or metal 'bucket'. As the net is pulled through the water the catch, or

Plate 17 A Smith–McIntyre grab being hauled back on-board

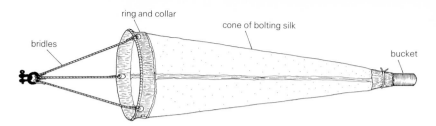

Fig. 7 A tow-net or plankton net

most of it, collects in the bucket and can be quickly removed when the net is brought to the surface. A modification of the simple tow-net, the Hensen net (fig. 8), has a canvas collar on the front of the net which restricts the size of the aperture. This has the advantage of reducing the volume of water flowing through the net and increasing the efficiency of filtration.

The netting traditionally used for plankton nets was bolting-silk but, although this material is still widely used, synthetic fibres such as nylon are becoming more popular as they do not rot after prolonged exposure

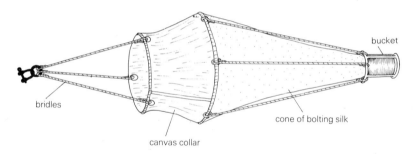

Fig. 8 A Hansen net

to sea water. The important feature of the netting is that it must retain a constant mesh-size under all working conditions, and bolting-silk was chosen originally because the fibres were specially woven to achieve this. When planning a plankton collecting trip it is important to give adequate consideration to the mesh size to be used, as this will determine the size of the organisms caught. Clearly, the finer the net the smaller the organisms retained, but also the greater the resistance of the tow-net to the passage of water and the quicker the net becomes clogged and ineffective. Simple tow-nets, consisting of a brass ring (300–450 mm in diameter), rope bridles, nylon netting in a variety of mesh

112

sizes from coarse (10 meshes per centimetre) to ultra-fine (240 meshes per centimetre), and metal or fibreglass buckets, can be obtained from the Marine Biological Association, Plymouth.

Simple nets such as those described above are adequate where general collections are required and where it is not important to know the volume of water filtered, or the operating depth of the net. However, if this sort of information is needed the basic plan can be modified in several ways. It is not possible to calculate the volume of water filtered by a plankton net simply from the mouth area of the net and the distance travelled, because no tow-net filters absolutely efficiently, and at least some of the water in front of the mouth is pushed to one side as overspill and never passes through the net. Since the filtering effi-

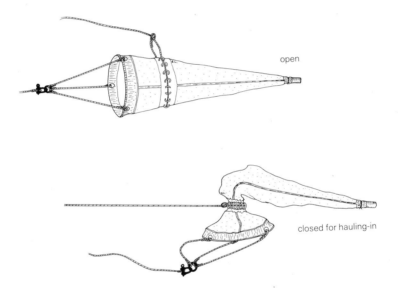

Fig. 9 A strangler device for a tow-net; a, open  b, closed for hauling-in

ciency of the net will depend on such factors as the towing speed, mesh size, and degree of clogging, the quantity of water that actually passes through the net can only be determined by measuring the flow with a flowmeter situated at some point inside the net. One of many such arrangements is described by Harvey (1934).

Similarly, the depth at which a net is fishing will depend on a variety of factors such as the form of the net itself, the towing speed, the amount of warp let out and the overall weight on the end of the warp. To have

113

exact knowledge of the working depth some kind of recording device is therefore necessary, either attached to the net itself or to the towing cable.

Finally, even if the path taken by the net is known, the catch will not have originated entirely from the main fishing depth since it will have been contaminated by organisms caught during the net's descent and ascent. To obtain a plankton collection from a discrete depth the net must be closed before hauling or ideally opened and closed at the required depth. The simplest type of closing mechanism is the strangler device, such as shown in fig. 9. When the net is to be hauled in, a messenger weight is sent down the warp to trip the mechanism and the strain is transferred to the strangler cable which pulls tight around the neck of the net. There are many more complex closing and opening devices in use operated by messenger weights, pressure release mechanisms, or by electronic signals from the surface, all designed to give more controlled sampling.

One problem common to all conventional plankton samplers is that they must be towed at very low speeds, usually no more than 1–2 knots. Even with nets 2 metres or more in diameter there is evidence that the larger and faster swimming planktonic animals are able to avoid them at such low towing speeds. A device widely used to collect these macroplanktonic organisms is the Isaacs-Kidd mid-water trawl. This is essentially a very large tow-net with a rectangular mouth-opening up to 3 metres or more across. It can be towed at relatively high speeds (4–5 knots) because the finer mesh netting is restricted to the cod end with the result that the net as a whole offers much less resistance to the flow of water than the more conventional nets. The trawl is prevented from rising in the water during the tow by a large V-shaped depressor plate mounted beneath the mouth.

Plankton samplers that will operate satisfactorily when towed behind commercial vessels at speeds of 10 knots or more have been developed (see for example Hardy, 1936; Gehringer, 1961) but they are beyond the scope of the general collector.

### (c) Freshwater habitats

The methods used for collecting animals from freshwater habitats are, in principle, similar to those described for the marine environment. The equipment differs mainly in size; the apparatus for use in fresh water being relatively small and lightweight.

Much of the collecting in shallow ponds and rivers will necessarily be carried out by hand or with the aid of a hand-net. The latter consists of a short net or bag made of canvas or other strong material attached to a rigid metal frame which is itself fixed to the end of a wooden handle. The length of the handle can be varied to suit the depth of water from which material is to be recovered. The construction must be very rigid to enable it to be dragged through vegetation and bottom debris, over rock surfaces, etc. On flat surfaces a D-shaped net is especially handy, as the flat side can be used as a scraper to remove sessile organisms and encrusting vegetation. Material collected with a hand-net is best emptied into a shallow dish for preliminary sorting, and the debris retained for closer examination with a hand-lens or a microscope.

A great variety of micro-organisms as well as larger invertebrates live amongst the submerged vegetation, and these need to be looked for very carefully. A lot of plant material can be collected by wading in shallow water, but a most useful tool for fishing up weeds from a boat or from the shore is the plant grapnel. This device consists simply of a cluster of metal prongs, weighted with lead to enable it to sink into the submerged vegetation and attached securely to a long line. An alternative to the grapnel is to use some sort of rake on the end of a long handle. The vegetation is first washed thoroughly with fresh water to remove the larger organisms, and then examined carefully under a microscope for sessile organisms. Some of the latter may be encouraged to leave the weed if it is left in a bowl of water for several hours to become stale. The more active rotifers and crustaceans will collect at the water surface while the less active micro-organisms such as tardigrades, sessile rotifers, some protozoans, gastrotrichs and some crustaceans will fall to the bottom. The organisms may possibly be concentrated to one side of the vessel by using a bright light to which they are attracted.

Samples from soft muddy deposits at the bottom of lakes and rivers can be obtained by using a simple corer, such as the F.B.A. automatic mud sampler. This corer consists of a Perspex tube about 500 mm in length and 50 mm in diameter fitted at the top with a metal valve. The device is lowered to the bottom with the valve open so that the tube sinks into the soft deposit, and as it is pulled out of the mud the valve closes automatically allowing the tube to withdraw a core of sediment. Suitable lead weights can be applied to the corer to achieve satisfactory penetration, but it cannot be said to be effective on gravel deposits or on weedy bottoms. If the water is no more than 4 metres deep the corer can be operated on the end of a wooden pole, in which case additional weights are not necessary. To extract the small organisms from the sample the sediment is shaken up with clean filtered water and allowed to settle, and the animals released from the soil particles are filtered from the supernatant liquid through a fine-mesh filter. The addition of an anaesthetic to the filtered water (magnesium chloride or formalin solution) may assist with the removal of the microfauna.

Damp sand on and above water level may contain a very specialized fauna including protozoa, nematodes, rotifers, tardigrades and harpacticoids. These organisms can be removed by agitation and filtration as outlined above.

For taking bottom samples in deep waters grabs or dredges similar to those described for collecting the marine benthic fauna (pp. 105–110) are most effective.

Planktonic animals from ponds and lakes can be collected quite simply with a conventional tow-net. This is made in just the same way as the plankton net used at sea; a conical fine-mesh net sewn to a canvas collar is attached to a brass ring about 300 mm in diameter, the narrow end of the net carrying a small collecting jar or funnel. Three ropes form the bridle of the tow-net which is tied to the towing line. If the net is being worked amongst rich vegetation it is often an advantage to have the mouth of the net covered by a cone of coarse wire gauze (Birge net) to prevent the bag from filling up with weeds. For work in shallow ponds an even smaller net is more suitable with a mouth opening of about 200 mm or less. As with all plankton nets the operating depth depends upon the speed at which it is pulled through the water and the weight on the end of the line. The number and size of animals filtered out by the net will depend on the mesh size incorporated, but some of the truly minute organisms such as some protozoans will pass through even the smallest mesh. These can only be concentrated by filtration through a membrane filter or by spinning the animals down in a centrifuge.

A piece of apparatus that is extremely useful for collecting in streams and rivers is the stream bottom sampler. In essence this is an open-ended metal box to which a conical net is attached. The mouth of the box is faced upstream so the current flows through the net. The bottom sediment in front of the box is stirred up thoroughly by hand and the organisms thus disturbed, together with a certain amount of debris, are washed into the net. From time to time the net is emptied into a collecting jar or shallow dish for examination.

## Parasitic animals

### (a) External parasites and other epizooic animals
As with all collecting it is important to remember that material must be properly labelled and this is never more necessary than with animal

parasites. Samples of parasites are of little value if the host from which they were taken is not carefully recorded. When handling animals that have been recently killed it is advisable to keep them separated as much as possible, as the external parasites tend to leave the body of the host as soon as it begins to cool down and it is important to know with which host-body the parasites were associated.

*Brushing and combing*  When dealing with dead hosts most external parasites can be quickly collected by simply combing and/or brushing the body over a shallow dish, although it is often advantageous to anaesthetize the parasites by first enclosing the host's body in a box containing a wad of cotton wool soaked in chloroform or ether. This method can also be used for collecting from many living animals, the host itself being anaesthetized if necessary.

*Other methods*  Unattached parasites in nests and soil can be collected by simple heat-desiccation methods (p. 00). Larger parasites such as ticks are best collected by direct examination of the host. However, it is essential to brush ticks with glycerol or alcohol before detaching them from the host. The ticks are relaxed by this treatment and they can then be removed without damaging the mouth parts. Unattached ticks can be collected from vegetation by a method known as 'flagging', in which a woollen blanket is trailed over the infested area.

In addition to the usual methods of direct searching, brushing and combing, a variety of procedures can be used for collecting ectoparasites from dead hosts. Fumigation of the bodies for about $1\frac{1}{2}$ hours in hydrogen cyanide* (generated from calcium cyanide) prior to brushing and combing has been recommended for collecting mammal ectoparasites. Bird parasites have been collected by fumigating the host's body in methyl bromide and then fluffing the feathers.

Lipovsky (1951) describes a method of removing ectoparasites from birds and mammals by washing the bodies, or in the case of larger mammals the pelt, in a detergent solution. This method has been found to be extremely effective for collecting fur and feather mites and larval Trombiculidae (chigger mites). Before washing, the body of the host should be refrigerated for at least one day; this procedure is evidently an important factor influencing the detachment of chigger mites from their hosts. The initial chilling is followed by a reasonable period of thawing or warming. The body is then placed in a jar of detergent

---

* Hydrogen cyanide is **extremely poisonous** and great care should be taken not to inhale the fumes.

solution and shaken from time to time over a period of about 30 minutes. The wash-detergent solution is then poured into a tall cylinder and allowed to settle for at least 15 minutes. The supernatant fluid is carefully decanted and the lower part poured into a shallow dish and examined under a binocular microscope for parasites.

Ectoparasites of mammals can also be collected by digesting the pelt of the host in the proteolytic enzyme trypsin, and then dissolving the remains in 10 per cent caustic potash solution (Cook, 1954). The skins are cut into pieces 30–50 mm square and placed in a flask containing a 3 per cent solution of trypsin buffered to about pH 8·3 with disodium phosphate solution. The flask is then incubated at 37°C for about 48 hours. Following digestion, 50 ml of a 20 per cent solution of caustic potash are added to the contents of the flask and the mixture boiled until all the hair and skin have dissolved. Finally, the solution is filtered through a sieve (100-mesh metal sieve is generally suitable) and the parasites washed off the sieve into a dish with alcohol. This method has been used successfully in quantitative surveys of mammal parasites, but has not been found to be satisfactory for bird ectoparasites.

Marine and freshwater fishes are often hosts to a variety of external parasites, belonging to three main groups, the monogeneans, copepods and isopods. These parasites are usually quite large and are easily visible to the naked eye. They can be removed from the host with a pair of fine forceps or by sliding a scalpel across the skin of the fish. However, some parasites may be more securely attached by means of hooks or powerful grasping limbs and must be removed more carefully to avoid damage. Dead fish often have a large amount of mucus on the body surface and this may conceal some of the smaller parasites. If the skin is scraped with a scalpel this mucus can be removed into a petri dish of clean water and examined closely.

The gills are another important site of parasite infection and should be examined carefully. The gills are best removed individually and examined in a dish of clean water. The gill parasites can be removed with forceps providing care is taken not to damage the attachment apparatus which is often embedded in the gill tissue. Other sites worth examining are the nostrils, the walls of the gill chamber and the mouth cavity.

Parasites collected from fishes are often contaminated with fish-mucus that should be removed before fixation. For this the parasites are placed in a glass-stoppered bottle containing sea water or fresh water as appropriate, and shaken vigorously. The water is then decanted off and the process repeated until the parasites are free from mucus.

In addition to aquatic vertebrate hosts, parasites are also found on representatives of most of the major groups of marine invertebrates. In many cases these parasites are copepod crustaceans. If a suitable host (for example, a starfish or a piece of sponge) is immersed in dilute sea water formalin (5 per cent) or alcohol (10 per cent) overnight, the associated animals become detached and can be collected from the bottom of the container the next morning. However, some of the parasites of invertebrates are internal and can only be collected after dissecting the host or teasing apart the tissues, as in the case of a sponge.

#### (b) Internal parasites

It is likely that every species of vertebrate may, under natural conditions, be infested with some internal parasites such as helminths or nematodes although they are not found in every individual nor will all parasitic species which attack a particular host be found in any individual of the host-species. Generally speaking, such parasites do not occur in suckling mammals and should be sought in animals which have passed this stage.

The positions which parasitic worms occupy are very varied. Many live in the alimentary canal or in its walls while many others invade other cavities and their lining membranes such as the liver, bile-ducts and gall-bladder, the nasal cavities and orbits, the lungs, trachea and bronchial tubes, the heart and blood vessels (especially the portal system), the kidneys, ureters, urinary bladder and the urethra. In birds the tough lining of the gizzard should be stripped off and examined for nematodes. In fishes, the swim-bladder and gill-chambers may also harbour these parasites.

A complete investigation of a host is a lengthy exercise and impossible under field-conditions. Examination has to be limited to the main natural cavities. Nevertheless, whether in the field or in the laboratory it is important to remember that only one host should be examined at a time.

When parasites are found in cysts or tumours in the walls of the alimentary canal, or imbedded in the tissues, it is advisable to cut out and preserve with them a portion of the surrounding tissue. Intestinal worms should be collected and preserved as soon as possible after the death of the host, as some of them deteriorate very rapidly, particularly those in fish-eating birds or in insectivorous mammals, and are then difficult to identify with any degree of certainty. To search for intestinal parasites, the gut of the host should be removed through an incision beneath the wing of birds or a slit in the abdomen of other vertebrates. The gut should then be slit open with scissors (preferably blunt-ended in order to avoid damage to the worms) and after the removal of its contents and any obvious parasites, the gut lining should be washed with normal saline solution and carefully examined for the less conspicuous parasites that may be attached to it. If salt solution is not available, plain water may be used, but it is liable to damage delicate worms, especially small threadworms, if they are left in it for more than a few minutes.

If the host is small, the whole gut may be opened in a dish of saline or water, or opened and then shaken up in a partly filled jar, but care should be taken as far as possible to note the region of the gut in which the parasites occur. Most adult cestodes and acanthocephalans are parasites of the intestine and burrow their heads deeply into the mucosa, often penetrating the muscular wall. In these instances it is advisable to leave the portion of tissue to which the head is attached in tap-water, as this will often induce the parasite to release its hold. Otherwise, the head will have to be dissected out with fine needles.

Some small worms are not easily detected unless the lining of the organ under examination is scraped with an instrument such as the back of a knife, and the scrapings shaken up in the washing fluid. This can then be turned out, a little at a time, into a large flat dish and the worms picked up using a needle or pipette, with the aid of a hand-lens or dissecting microscope if necessary. An alternative plan is to shake the material up in a tall jar, and allow it to stand for a few minutes. The worms will sink to the bottom, and most of the dirty liquid can be decanted off. Clean fluid is then added, and the process repeated until the worms are comparatively free of debris.

Trematodes and nematodes occur also in the lungs, the liver and the kidneys. These organs should be placed in a normal saline solution and broken up with a pair of strong needles. The worms will be found in the fluid or amongst the broken tissue, or they may emerge from within the tissues if the pieces are gently pressed.

Specimens should be placed in small tubes filled with preservative (cotton wool must not be used to pack the tubes) together with the appropriate labels. The label should at least bear the name of the host or the same reference number as that of the host, as well as the precise location of the parasites within the body, the locality of the host and the date of collection.

Facilities may not always be available for the early helminthological examination of the hosts, and in such cases the body may be preserved whole for study at a later date. When this situation arises it is most important that an incision should be made in the host's abdomen and preserving fluid injected as soon as possible. Generally, however, helminths preserved in this manner are found to be in very poor condition.

## References

Anderson, R. O. 1959. A modified flotation technique for sorting bottom fauna samples. *Limnol. Oceanogr.* **4,** 223–225.

Birkett, L. 1957. Flotation technique for sorting grab samples. *J. Cons. perm. int. Explor. Mer* **22,** 289–292.

Boisseau, J.-P. 1957. Technique pour l'étude quantitative de la faune interstitielle des sables. *C.R. Congr. Socs sav. Paris Sect. Sci.* 117–119.

Cook, E. F. 1954. A modification of Hopkin's technique for collecting ectoparasites from mammalian skins. *Ent. News* **65,** 35–37.

Forster, G. R. 1953. A new dredge for collecting burrowing animals. *J. mar. biol. Ass. U.K.* **32,** 193–198.

Gehringer, J. W. 1961. The Gulf III and other modern high-speed plankton samplers. *Rapp. Cons. Explor. Mer* **153,** 232 pp.

Hardy, A. C. 1936. The continuous plankton recorder. *Discovery Rept.* **11,** 457–510.

Harvey, H. W. 1934. Measurement of phytoplankton population. *J. mar. biol. Ass. U.K.* **19,** 761–773.

Higgins, R. P. 1964. A method for meiobenthic invertebrate collection. *Amer. Zool.* **4,** 291.

Hiscock, K. & Hoare, R. 1973. A portable suction sampler for rock epibiota. *Helgoländer wiss. Meeresunters* **25,** 35–38.

Holme, N. A. 1961. The bottom fauna of the English Channel. *J. mar. biol. Ass. U.K.* **41,** 397–461.

Holme, N. A. 1964. Methods of sampling the benthos. *Adv. mar. biol.* **2,** 171–260.

Holme, N. A. & McIntyre, A. D. 1971. Methods for the study of marine benthos. *I.B.P. Handbook* No. 16. Oxford and Edinburgh: Blackwell Scientific Publications. 334 pp.

Hulings, N. C., & Gray, J. S. 1971. A manual for the study of meiofauna. *Smithson. contr. Zool.* **78,** 84 pp.

Lipovsky, L. J. 1951. A washing method of ectoparasite recovery with particular reference to chiggers (Acarina: Trombiculidae). *J. Kansas ent. Soc.* **24,** 151–156.

Moore, H. B., & Neill, R. G. 1930. An instrument for sampling marine muds. *J. mar. biol. Ass. U.K.* **16,** 589–594.

Murphy, P. W. 1962. Extraction methods for soil animals. In Murphy, P. W. (Ed.) *Progress in Soil Zoology.* London: Butterworths. pp. 75–114.

Raw, F. 1955. A flotation extraction process for soil micro-arthropods. In Kevan, D. K. McE.(Ed.) *Soil Zoology*. London: Butterworths. pp. 341–346.

Smith, W., & McIntyre, A. D. 1954. A spring-loaded bottom sampler. *J. mar. biol. Ass. U.K.* **33,** 257–264.

Uhlig, G. 1964. Eine einfache Methode zur Extraktion der vagilen, mesopsammalen Mikrofauna. *Helgoländer wissenschaftliche Meeresuntersuchungen* **11,** 178–185.

# Chapter 3

# Killing, fixing and preserving

The killing and preservation of some invertebrates present no great problems as the animals can be dropped directly into a suitable preserving fluid – usually strong alcohol or a solution of formaldehyde not weaker than 3 per cent. However, depending on the group and on the purpose for which the material is required, some pre-preservation treatment, namely anaesthetization (or narcotization) and/or fixation may be necessary.

### Anaesthetization
Many invertebrates are highly contractile and if they are to be preserved in a relaxed condition they must first be slowly anaesthetized either until dead or until they are so insensitive that they can be killed in an extended condition with fixative or preservative fluids. In general, animals should be anaesthetized for as short a period of time as possible, for if the anaesthetization is prolonged tissue breakdown may begin, although such changes would be unimportant if the material is to be preserved for its gross interest rather than for micro-anatomical or cytological study.

## Menthol

This has been recommended for difficult sessile animals such as Polyzoa. The animals should be allowed to expand in clean water (sea water for marine species) and crystals of menthol scattered on the surface. The time required for anaesthetization may be up to 12 hours.

## Magnesium sulphate

Magnesium sulphate is sometimes used as a saturated solution into which the animals are plunged, but more satisfactory results are obtained if the crystals are sprinkled gradually on to the surface of the water containing the animals. Alternatively, magnesium sulphate can be introduced gradually into the water over a period of some hours in the form of a 20–30 per cent solution. Magnesium sulphate has given satisfactory results with a great variety of animals such as giant nudibranchs, chitons, madreporaria and some alcyonarians, but a general drawback to its use is that about 150g per litre of sea water is required, and the long duration of treatment and the osmotic pressure developed is likely to cause changes in the tissues.

## Magnesium chloride

A nearly isotonic solution of magnesium chloride (about 7·5 per cent $MgCl_2.6H_2O$ dissolved in fresh or distilled water) has been widely used for anaesthetizing marine animals. Small planktonic animals such as medusae can be placed into a watch-glass with this solution and left for half a minute or so before being transferred by pipette to formalin. Larger animals will have to be transferred to vessels of magnesium chloride which can then be poured off and replaced by formalin.

## Chloral hydrate

This substance has also been used successfully with a wide range of marine and freshwater animals. The crystals are sprinkled on the surface of the water containing the animals, or alternatively the animals can be placed directly in fresh solutions of up to about 2 per cent in strength.

## MS 222-Sandoz

This is the manufacturer's code name for ethyl m-aminobenzoate, a compound which has been widely used as an anaesthetic for cold-blooded vertebrates and which has been shown to be useful for certain invertebrates. For example 0·01–0·02 per cent aqueous solutions have been used very successfully for anaesthetizing Malacostraca (Crustacea) of all sizes. Treatment time varies from a few seconds to 10 or 15 minutes. MS 222 is manufactured by Sandoz Products Ltd, London.

## Propylene phenoxetol

In addition to being used as a post-fixation preservative (p. 135) this substance has been used successfully as an anaesthetic for oligochaetes, various molluscs, malacostracan Crustacea and spiders, and it will almost certainly prove to be a useful anaesthetic for most invertebrate animals. Animals can be immersed in clean sea water or fresh water (as appropriate) and propylene phenoxetol added so that a large globule of the viscous compound forms at the bottom of the container. The amount of propylene phenoxetol added should not exceed 1 per cent of the volume of water in the container. The time required for anaesthetization varies. Treatment for several hours has been required to render insensitive clams (*Tridacna*) 120–150 mm in length, oligochaetes on the other hand have been sufficiently anaesthetized after treatment for 10–15 minutes. An additional benefit is that animals appear to recover from relatively long periods of exposure to the anaesthetic when returned to clean water. It has been reported that specimens of *Tridacna* which had been anaesthetized for over six hours seemed normal after being returned to running sea water overnight. Propylene phenoxetol is manufactured by Nipa Laboratories Ltd, Treforest, Glamorganshire, Great Britain.

## Ethyl alcohol

Ten per cent ethyl alcohol made up from absolute alcohol and not from Industrial Methylated Spirit (see p. 133) has been recommended as an anaesthetic for freshwater animals. A small quantity of the 10 per cent alcohol should be added to the water and excitation in the animals allowed to subside before more is added. The time required for anaesthetization varies from a few minutes to about an hour.

## Benzamine hydrochloride/cellosolve mixture

This has been used successfully for the anaesthetization of rotifers and is prepared as follows:

| | |
|---|---|
| Benzamine hydrochloride, 2 per cent aqueous | 3 parts |
| Cellosolve, pure (ethylene glycol monoethyl-ether) | 1 part |
| Distilled water | 6 parts |

## Eucaine (β-eucaine hydrochloride)

A solution of eucaine may be used for anaesthetizing small aquatic animals such as rotifers and coelenterates. It is made up as follows:

| | |
|---|---|
| Eucaine | 1 g |
| Alcohol (90 per cent) | 10 ml |
| Distilled water | 10 ml |

The solution should be added gradually to the water containing the animals.

## Stovaine (Amyl chlorohydrin)

This substance may also be used as an anaesthetic for small animals. For preparation, stovaine may be substituted for eucaine in the above formulation.

## Other substances

Many other substances can be used as anaesthetics. Tobacco smoke is effective for many small organisms including Hydroida and ciliates. It should be slowly bubbled into the water through a fine glass tube resting on the bottom of the container. Carbonic acid gas ($CO_2$) introduced by squirting soda water from a siphon into water containing the animals has been used successfully for many Coelenterata and Echinodermata. Ether, chloroform or ethyl acetate vapour is sometimes used for anaesthetizing and killing terrestrial arthropods although very often these animals require no special treatment before preservation. Animals can be treated by being enclosed in a tube with cotton wool to which a few drops of the liquid have been added.

**Fixation**
Fixation is a process which stabilizes the protein constituents of tissue so that after death or even after treatment such as embedding, sectioning and mounting, the tissue constituents retain in some degree the form they possessed in life. Additionally, fixation raises the refractive index of the cell contents and renders tissue more easily stainable. Most chemical fixatives coagulate protein although some, notably formaldehyde, act by making the protein sols more viscous or by converting them into gels. While some fixatives are more generally useful than others, no one fixative can be regarded as an all-purpose formulation and the choice of fixative must depend on the material to be fixed and on the purpose for which the fixed material is required. The substances most commonly used as fixatives are formaldehyde, ethanol (ethyl alcohol), acetic acid, picric acid, mercuric chloride, osmium tetroxide (osmic acid), potassium dichromate and chromium trioxide (chromic acid). Some of these compounds may be used alone, but more often mixtures are made, the object being to combine the virtues of the various ingredients. Some of the more widely used formulations are listed

below, but a variety of problems can arise in connexion with fixation and if material is to be collected primarily for histological or cytological study collectors should seek expert advice or consult specialist texts such as Pantin (1969) and Gatenby & Beams (1950).

## Formaldehyde

Commercial formalin is usually purchased as a 40 per cent aqueous solution of the gas formaldehyde, together with various impurities. Diluted to a 10 per cent formalin solution it is a useful cytoplasmic fixative. In making dilutions it is important to keep in mind the difference between a given percentage of formaldehyde and a given percentage of formalin solution. To make 10 per cent *formaldehyde* one adds 3·5 parts of water to 1 part commercial formalin (40 per cent formaldehyde), while to make 10 per cent *formalin solution* one adds 9 parts water to 1 part commercial formalin. To prevent distortion following osmotic changes, for marine animals the 10 per cent formalin solution should be made up with sea water. Formaldehyde solutions almost always have an acid reaction due to the presence of formic and other acids and are thus unsuitable for treating calcareous animals. However, formalin is one of the few penetrating fixatives which can be used as a neutral solution. Solutions can be neutralized in several ways. For example a 4 per cent solution of sodium hydroxide can be added drop by drop until the formalin is neutral to phenol red. Alternatively the formalin can be buffered with the organic base hexamine (hexamethylenetetramine), the appropriate concentration being 200 g of hexamine per litre of undiluted commercial formalin. Formaldehyde solutions often become turbid on standing due to the production of paraformaldehyde, but according to some authorities this can be avoided by storing solutions in darkened bottles in a cool store. Animals are generally fixed in formalin for 48 hours.

## Steedman's solution

This mixture is recommended for general fixation and preservation of marine zooplankton (Steedman, 1976). It has the advantage of combining the fixative properties of formalin with the preservative and softening action of propylene phenoxetol and propylene glycol. The solution is prepared as follows:

| | |
|---|---:|
| Propylene phenoxetol | 0·5 ml |
| Propylene glycol | 4·5 ml |

| Formalin solution, commercial 40 per cent | 5 ml |
| Sea water (or distilled) | 90 ml |

alternatively a strong stock solution can be prepared

| Propylene phenoxetol | 50 ml |
| Propylene glycol | 450 ml |
| Formalin solution, commercial | 500 ml |

and diluted 10 ml stock solution with 90 ml sea water for general use.


## Bouin's fluid (Picro-Formol)

This is an excellent micro-anatomical fixative for marine invertebrates. It is made up as follows:

| Picric acid, saturated aqueous solution* | 75 ml |
| Formalin (commercial | 25 ml |
| Acetic acid (glacial) | 5 ml |

Material is fixed for at least 12 hours, but many animals may be left in Bouin's indefinitely.


## Alcoholic Bouin (Dubosq–Brasil fluid)

This is a more penetrating fixative than Bouin's fluid and hence is particularly suitable for animals with a tough exoskeleton such as the arthropods. Its composition is:

| Picric acid* | 1 g |
| Acetic acid (glacial) | 15 ml |
| Formalin (commercial) | 60 ml |
| Alcohol (80 per cent) | 150 ml |

Fixation time is about two hours – perhaps rather longer for large and heavily sclerotized specimens, but if the fixation time is greatly prolonged the material tends to become brittle.

---

* This substance is explosive and detonates readily when in contact with certain metals. It should therefore never be kept in a metal container. The safest way to store it is in a glass vessel under water.

# Heidenhain's Susa mixture

This is probably the most generally useful fixative for material which is to be sectioned. Gross cell structure is well preserved but as with other fixatives containing acetic acid fine cytoplasmic detail is lost. It has the composition:

| | |
|---|---|
| Mercuric chloride* | 45 g |
| Sodium chloride | 5 g |
| Distilled water | 800 ml |
| Trichloracetic acid | 20 ml |
| Acetic acid (glacial) | 40 ml |
| Formalin (commercial) | 200 ml |

The first three items are generally made up as a stock solution and for marine animals better results may be obtained by substituting sea water for distilled water. This fluid and other solutions containing mercuric chloride must not be touched with metals as these can produce precipitates which damage tissue. Specimens can be manipulated in the solution with glass rods or wooden spills. Fixation time is 3–24 hours and, in order to remove mercuric precipitates from the tissue, after fixation material must be transferred to 90 per cent iodized alcohol (p. 131).

# Viets' solution

| | |
|---|---|
| Glacial acetic acid | 3 parts |
| Glycerin | 11 parts |
| Distilled water | 6 parts |

This solution is recommended for the fixation and preservation of water mites.

# Oudemans' fluid

| | |
|---|---|
| Glacial acetic acid | 8 parts |
| Glycerin | 5 parts |
| Alcohol, 70 per cent | 87 parts |

This solution is favoured by some workers for fixing and preserving terrestrial mites.

---

* See footnote on mercuric chloride on p. 130.

## Zenker's fluid

This is another useful fixative for micro-anatomical work. Although it preserves fine cytoplasmic structures better than 'Susa', a rather more elaborate post-fixation treatment is required. The composition of the fluid is:

| | |
|---|---|
| Mercuric chloride* | 5 g |
| Acetic acid (glacial) | 5 ml |
| Potassium dichromate | 2 g |
| Sodium sulphate | 1 g |
| Distilled water | 100 ml |

Since it contains both oxidizing and reducing substances the fluid does not keep well, and as in the case of 'Susa' the fixative should not come in contact with metal. Fixation time is 3–12 hours and, in order to remove mercuric precipitates, after fixation the material must be thoroughly washed in running or frequently changed water. After washing, the material is transferred to 50 per cent alcohol.

## Flemming's solution

A useful fixative for small invertebrates, prepared as follows:

| | |
|---|---|
| Chromic acid, 1 per cent | 150 ml |
| Osmic acid, 2 per cent† | 40 ml |
| Acetic acid, glacial | 10 ml |

After fixation, the material should be washed in running water to remove traces of osmic acid which might otherwise cause blackening.

## Chromic/Osmic acid mixture

For use as a fixative it should be made up as follows:

| | |
|---|---|
| Chromic acid, 1 per cent | 100 ml |
| Osmic acid, 1 per cent† | 2 ml |

---

* *Important* Corrosive sublimate (mercuric chloride) is extremely poisonous, and the saturated solution looks just like plain water and has no odour. Also, it should not be brought into contact with steel instruments since it has a strong corrosive reaction.
† *Important* Osmic acid must be handled with care because the solution and vapour are highly toxic.

## Chromic/Acetic acid mixture

For use as a fixative it should be made up as follows:

Chromic acid, 1 per cent      100 ml
Acetic acid, glacial           5 ml

## Corrosive sublimate (mercuric chloride)*

This substance is used as a fixative, usually as a saturated solution, in either fresh water or sea water as appropriate. Because of the *extremely* poisonous nature of this substance it is not used very widely and is best avoided where possible. After fixation it is most important that all traces of the solution are removed from the specimens or the material will become spoilt. The simplest method is to wash for several hours in running water but where this is impracticable the specimens may be washed in numerous changes of water or 70 per cent alcohol. The most certain method of removing all traces of mercuric chloride is to soak the material overnight in iodized alcohol (see below).

## Corrosive acetic

The addition of a small quantity (from a few drops up to 10 per cent) of glacial acetic acid to corrosive sublimate fixative may reduce the tendency for shrinkage of delicate tissues.

## Iodized alcohol

This reagent is used to remove traces of corrosive sublimate from fixed material. The solution is prepared as follows:

Iodine                      3 g
Potassium iodide       6 g
Alcohol, 70 per cent    300 ml
This solution, known as tincture of iodine, is then added to 70 per cent alcohol in sufficient quantity to give a brown sherry colour. The specimens are washed in the iodized alcohol until the brown colouration no longer disappears, and then transferred to fresh 70 per cent alcohol to remove any traces of iodine which might remain in the material.

---

* See footnote on mercuric chloride on p. 130.

## Schaudinn's solution

This is an alcoholic solution of corrosive sublimate (saturated mercuric chloride) made up in the following manner:

| | |
|---|---|
| Saturated mercuric chloride* | 2 parts |
| Alcohol, 90 per cent | 1 part |

A few drops of glacial acetic acid may be added before use to reduce the tendency for shrinkage of delicate tissues.

## T.A.F.

This fixative is employed in some instances for the fixation of nematodes. The reagent is prepared as follows:

| | |
|---|---|
| Formalin, commercial | 14 ml |
| Triethanolamine | 4 ml |
| Distilled water | 82 ml |

## Dowicil 100

Dowicil 100 is the proprietary name for 1-(3-chlorallyl)5,7-triaza-1-azoniaadamantane chloride. It is a yellow/white solid, highly soluble in water and as a biological fixative is used as a 10 per cent aqueous solution (fresh water or sea water as appropriate). Its action as a fixative is reported to depend on its releasing formaldehyde only in the presence of proteins. As with formalin, the slightly acid solution can be neutralized with limestone or marble chips without affecting the preservative action. Specimens placed in Dowicil solution retain their shape and remain quite flexible, showing none of the brittleness often associated with formalin-preserved material. As well as a fixative Dowicil can also be used as a preservative. It has advantages over formalin because it does not give off irritative fumes or cause skin disorders, and over alcohol in that it is non-flammable and non-volatile. In comparison with formalin and industrial spirit, Dowicil may be rather expensive, but because of the difficulty and high cost of transporting liquid preservatives it could prove to be a convenient and economical fixative/preservative for expedition work.

---

* See footnote for mercuric chloride on p. 130.

**Preservation**

The processes of fixation and preservation are often confused. Essentially a preservative is a fluid in which material can be stored indefinitely, which, without seriously distorting specimens or destroying their constituent parts, arrests autolysis of cells (i.e. the self-digestion of cells by the enzymes present within them) and which also destroys bacteria and moulds. The arrest of autolysis is regarded by some as part of the fixation process but until recently this distinction was in a sense academic as all of the commonly used preservatives had some fixative action. However, with the introduction of 'phenoxetols' (p. 135) the distinction becomes important. These materials do not arrest autolysis and, therefore, they must either be regarded as 'incomplete preservatives' or as *post-fixation* preservatives.

## Formaldehyde

Used as a 5–10 per cent aqueous solution formalin is a good general preservative. Preserving solutions can be made up by diluting buffered commercial formalin (see p. 127) with fresh water or sea water as appropriate. Formaldehyde has the advantage of being cheap and non-flammable but it tends to stiffen and harden animals so that they become brittle and difficult to handle. Because of its pungent vapour it can be distinctly unpleasant to use.

## Dowicil

As well as a useful fixative solution (p. 132) 10 per cent agueous Dowicil can be used as an effective preservative.

## Ethyl alcohol (Ethanol)

Ethyl alcohol diluted with water to a concentration of about 70 per cent by volume has for many years been regarded as the best general preservative for invertebrate animals. Collectors need not assess the strength of the alcohol used with minute accuracy, but it should be ensured that the strength of the spirit in which specimens are finally stored is at least 50 per cent. In estimating the strength of the final preservative it should be remembered that alcohol extracts water from specimens placed in it, and, since animal tissue is made up largely of water, if the bulk of alcohol does not exceed that of the fresh specimens placed in it, its strength will eventually be reduced by at least one half.

Provided an adequate bulk of fluid is used many small animals with tough exoskeletons can be placed directly in 70 per cent spirit, but with soft-bodied animals direct transference to strong alcohol is apt to cause shrinkage. This can be avoided by first placing them in weaker solutions and gradually increasing the strength. Thus soft-bodied animals should ideally be placed for a few hours in 30 per cent alcohol and then transferred for a similar period of time to 50 per cent alcohol. Thereafter they should be left for a few days in 70 per cent alcohol before being finally stored in clean spirit of the same strength. The strength of the final preservative can be increased but many animals become unduly hard if stored in alcohol stronger than 80 per cent.

In Britain, and indeed in most other countries, pure ethanol is very expensive owing to the high tax which has to be paid, and for routine preservation ethanol in the form of methylated spirit is used. Methylated spirit is a mixture of ethyl alcohol and either crude or pure methyl alcohol. In Britain two *broad* categories of methylated spirit are available, mineralized and non-mineralized. Mineralized spirit, some forms of which are dyed purple or red, contains a small amount of pyridine and the methyl component is crude methyl alcohol (wood naphtha). This type of methylated spirit is quite useless as a preservative since when diluted with water it forms a milky emulsion. Non-mineralized spirit consists of 95 parts by volume of ethanol and 5 parts by volume of either crude or pure methyl alcohol, depending on the grade. The grade of non-mineralized spirit called industrial methylated spirit (IMS) is quite suitable for preserving biological material and its qualities are listed in Notice No. 58 of the Commissioners of Customs and Excise (London: H.M.S.O.).

IMS can be bought in small quantities from retail chemists or in larger quantities from methylators, but purchases can only be made by holders of permits issued by the Commissioners of Customs and Excise. *Bona fide* collectors should have no difficulty in obtaining the necessary permits and application should be made to the nearest office of the Customs and Excise.

The Sikes hydrometer used by excise officers in Britain indicates the strength of alcohol in relation to what is termed 'proof spirit', a mixture composed of about equal parts by weight of alcohol and water. Strengths of alcohol weaker than proof spirit are measured on a scale of 100 degrees while those stronger are measured on a scale of about 75 degrees. Thus pure water is 0 per cent proof ($100°$ under proof) and pure spirit is 175·35 per cent proof ($75·35°$ over proof). The relationship between British proof strengths and certain other scales is given in Table I.

134

Table I. The relationship between specific gravity at 15·6°C, British proof strengths and percentages of alcohol by weight and volume.

| Specific gravity | Degrees proof | Per cent alcohol | |
|---|---|---|---|
| | | By weight | By volume |
| 1·0000 | 0 | 0 | 0 |
| 0·9928 | 8·77 | 4·02 | 5·03 |
| 0·9866 | 17·44 | 8·04 | 9·99 |
| 0·9811 | 26·15 | 12·12 | 14·98 |
| 0·9760 | 34·87 | 16·25 | 19·98 |
| 0·9709 | 43·64 | 20·42 | 24·98 |
| 0·9654 | 52·44 | 24·67 | 30·00 |
| 0·9591 | 61·20 | 28·97 | 35·00 |
| 0·9518 | 69·98 | 33·36 | 40·00 |
| 0·9435 | 78·77 | 37·86 | 45·00 |
| 0·9343 | 87·53 | 42·48 | 49·99 |
| 0·9242 | 96·32 | 47·24 | 55·00 |
| 0·9198 | 100·00 | 49·28 | 57·10 |
| 0·9134 | 105·10 | 52·15 | 60·01 |
| 0·9020 | 113·84 | 57·18 | 64·98 |
| 0·8899 | 122·63 | 62·42 | 69·99 |
| 0·8772 | 131·39 | 67·85 | 74·98 |
| 0·8637 | 140·21 | 73·51 | 80·00 |
| 0·8494 | 148·98 | 79·41 | 84·98 |
| 0·8337 | 157·79 | 85·68 | 90·01 |
| 0·8159 | 166·56 | 92·40 | 95·00 |
| 0·7936 | 175·35 | 100·00 | 100·00 |

## Propylene phenoxetol

As a 1–2 per cent aqueous solution this substance has been used successfully as a post-fixation preservative for a wide range of vertebrate and invertebrate animals. It has also been reported to preserve well the natural colour of specimens and leaves the material pliable. The one per cent solution is more expensive than 80 per cent commercial methylated spirit, but it has the advantages of being non-flammable and non-volatile. In many institutions it is now replacing spirit as a standard preservative, and for expedition work has the obvious advantage that only small quantities of the substance (a viscous fluid) need be transported to make up solutions in the field. It must be emphasized that propylene phenoxetol should only be used after adequate fixation. If fresh unfixed material is stored in the solution it will almost certainly decompose.

Propylene phenoxetol is extremely difficult to dissolve in water and requires very vigorous stirring to obtain a solution of one per cent by volume. Steedman (quoted by Cooke 1969) has recommended the use of propylene glycol as a coupler to facilitate the preparation of phenoxetol solutions. The stock solution is prepared by mixing 20 ml of propylene phenoxetol with 50 ml of propylene glycol (referred to as phenoxypropylene glycol or PPG). The preserving solution is then prepared by taking 7 ml of stock solution and making up to 100 ml with water and shaking briefly. This gives a 2 per cent phenoxetol and 5 per cent glycol mixture, the glycol offering the advantage of being a powerful fungicide and further restricting evaporation. It has also a low viscosity and is therefore not sticky and is the only glycol non-toxic to man. It has been pointed out that the penetration of PPG into tissues is rather poor and slow, but as it is readily miscible with alcohol which does penetrate freely, and immersion of the material in alcohol for 24 hours before transferring to the preservative may be worthwhile.

## Phenoxetol BPC

As a 1–2 per cent aqueous solution Phenoxetol BPC (B-phenoxyethylalcohol) has also been used as a post-fixation preservative. It is rather cheaper than propylene phenoxetol but apparently it is less efficient as a bactericide and much less efficient as a fungicide. Phenoxetol BPC and propylene phenoxetol are manufactured by Nipa Laboratories Ltd, Treforest, Glamorganshire, Great Britain.

## Ethylene glycol

Used as a 50 per cent aqueous solution this substance (probably better known as radiator antifreeze) is reported to be a useful preservative for marine organisms. It has the advantage over ethyl alcohol that it is non-volatile, non-flammable and does not precipitate when mixed with sea water or fresh water with a high mineral content. With marine plankton the best results were obtained by fixing the material with formalin, decanting and washing in sea water, followed by the addition of ethylene glycol. It is said to cause little or no shrinkage of the small planktonic organisms.

*References*

Cooke, J. A. L. 1969. Notes on some useful arachnological techniques. *Bull. Brit. arach. Soc.* **1,** 42–43.

Gatenby, J. B., & Beams, H. W. 1950. (Edit.) *The Microtomist's Vade-Mecum* (Boles Lee). London: Churchill. 753 pp.

Grimstone, A. V., & Skaer, R. J. 1972. *A Guidebook to Microscopical Methods.* Cambridge University Press. 134 pp.

Pantin, C. F. A. 1969. *Notes on Microscopical Technique for Zoologists.* Cambridge University Press. 77 pp.

Steedman, H. F. 1976. *Zooplankton Fixation and Preservation.* Paris: The Unesco Press. 350 pp.

Chapter 4

# General treatment of collections

**Labelling**

Material without any data is practically useless for most kinds of taxonomic research. At the very least specimen labels should carry a note of the locality, the date of collection and the name of the collector. It may often be desirable to define the locality in some detail, for in the case of some animals – for example terrestrial gastropods – racially distinct populations may occur within very short distances of each other. Generally the country, county or state, grid reference, or, if at sea, latitude and longitude, should be recorded. If the locality is a small geographical feature its position in relation to a well-defined spot should also be noted, for example 'small creek on W side of Hille track 2·5 km N of Dhankuta'.

Ecological data are often of great value, and indeed essential in the case of phytophagus species, parasites and other species living in close association with other animals. The sort of information that should be recorded will depend to some extent on the animals collected. For free-living animals a brief description of the habitat should always be recorded, and this should include for example the nature of the bottom for benthic animals and the type of substratum for sedentary species. Vague general description should be avoided. Clearly 'moss at foot of oak tree' is more informative than 'woodland'. Data on depth should be recorded for aquatic species, and for terrestrial species collected in hilly country it is useful to record the altitude.

When dealing with external parasites, other epizooic species, and internal parasites, the name of the host animals should always be recorded. If the host cannot be precisely determined, the collector should note as nearly as possible the kind of animal involved, for example, 'frog', 'mouse', 'starfish', 'sponge' and so on. Alternatively the host itself can be preserved and a cross reference made on the labels. With external parasites and other epizooic species the position on the host should be recorded, and with internal parasites a note should be made of the organ or possibly the tissue in which they were found. Similarly with phytophagus species the name of the host plant should be noted or a cross referenced specimen preserved for subsequent determination.

For large-scale collecting expeditions it is generally worth while to have special pre-printed labels. These can be made in sizes suitable for the various tubes and bottles to be used. Labels of this sort have several advantages – they are time saving, legibility is increased, and the collector is reminded of the data that should be given. Examples of types of labels suitable for marine and terrestrial expeditions are shown below. The use of a station number (St. No.) on the marine label instead of giving habitat data is a useful economy. This number refers to a master list containing full details relevant to the collecting operations. Although the use of station numbers is generally confined to marine collecting, the method could be useful for certain terrestrial expeditions.

| | |
|---|---|
| NEW GUINEA: MOROBE DISTRICT; Edie Creek *c.* 6 miles SW of Wau.<br><br><br>M.E. Bacchus *Coll. No.* , . .1964<br>British Museum (Nat. Hist.) and Univ.<br>of Newcastle-upon-Tyne Exped. 1964-65 | Reg. no.:<br><br>ex<br><br>S.E. KENYA:<br><br><br>B.M.(N.H.) Exped. to East Africa 1965 |
| MUKTINATH 28°48.5'N 83°52.5 E<br>Northern slopes of Muktinath Himal<br>13,000 ft.<br>K. H. HYATT. *Coll. No.*<br>Brit. Mus. Nepal Exped. 1954. | Loc.<br><br><br>*Coll.* *Det.* |
| Milke Bhanjyang (27°19'N 87°31'E)<br>Rhododendron forest<br>10–11,000 ft. *Coll. No.*<br>Brit. Mus. Nepal Exped. 1961–1962. | |

More detailed ecological and other information, including for example data on temperature, salinity, frequence of occurrence, associated species, colour of the living animals and so on, can be recorded in field notebooks. However, essential data should always be recorded on labels since notebooks may be overlooked and could be unavailable if the collection is divided.

When collections are being made in association with ecological surveys, it may be convenient to record the data on specially prepared field record cards. Cards of this sort have been designed and prepared by the Biological Records Centre, Monks Wood Experimental Station, Huntingdon, Great Britain, an organization responsible for collecting data on the distribution of much of the British flora and fauna. The card illustrated on page 142 is used to record general data for

| | | | LOCALITY | | NON-MARINE ISOPODS 6567 (1–4) | |
|---|---|---|---|---|---|---|

(22–24)

**NON-MARINE ISOPODS 6567** (1–4)

| Date | (60–64) | V.C. No. | (33–35) |
|---|---|---|---|

| RECORDER | | DET. | (77–79) V.C. |
|---|---|---|---|

| | Alt. | (73–76) | Code No. (65–68) |
|---|---|---|---|

Grid Ref. (25–32)

| (5–10) | | | (5–10) | | |
|---|---|---|---|---|---|
| 00101 | Acaeroplastes melanurus | | 07601 | Metoponorthus cingendus | |
| 00501 | Agabiformius lentus* | | 07602 | pruinosus | |
| 00901 | Androniscus dentiger | | 07701 | Miktoniscus linearis* | |
| 01001 | Armadillidium album | | 08301 | Nagurus cristatus* | |
| 01002 | depressum | | 08302 | nanus* | |
| 01003 | nasatum | | 08801 | Oniscus asellus | |
| 01004 | pictum | | 08901 | Oritoniscus flavus | |
| 01005 | pulchellum | | 09401 | Philoscia muscorum | |
| 01006 | vulgare | | 09601 | Platyarthrus hoffmannseggi | |
| 01301 | Asellus aquaticus | | 09701 | Porcellio dilatatus | |
| 01302 | cavaticus | | 09702 | laevis | |
| 01303 | communis* | | 09703 | scaber | |
| 01304 | meridianus | | 09704 | spinicornis | |
| 03201 | Chaetophiloscia meeusei* | | 10201 | Reductoniscus costulatus* | |
| 03202 | patiencei* | | 12001 | Trachelipus rathkei | |
| 03301 | Cordioniscus spinosus* | | 12002 | ratzeburgi | |
| 03302 | stebbingi* | | 12101 | Trichoniscoides albidus | |
| 03801 | Cylisticus convexus | | 12102 | saeroeensis | |
| 04701 | Eluma purpurascens | | 12103 | sarsi | |
| 05201 | Halophiloscia couchi | | 12201 | Trichoniscus pusillus agg. | |
| 05401 | Haplophthalmus danicus | | 12201/1 | provisorius | |
| 05402 | mengei | | 12201/2 | pusillus | |
| 06801 | Ligia oceanica | | 12202 | pygmaeus | |
| 06901 | Ligidium hypnorum | | 12401 | Trichorina tomentosa* | |

*Species probably alien

Other species:

## HABITAT DATA

**A** 1 tick (obligatory):

| | 11 |
|---|---|
| Coastal <15km from sea | 1 |
| Inland >15km from sea | 2 |

**B.** 1 tick (obligatory):

| | 12 |
|---|---|
| Urban | 1 |
| Suburban/village | 2 |
| Rural | 3 |

**C** 1st order habitats;
1 tick (obligatory):

| | 13–15 |
|---|---|
| Aquatic: Canal | 001 |
| River >5m wide | 002 |
| Lake >1 acre (0.4 hectare) | 003 |
| Estuary | 004 |
| Sea | 005 |
| Marsh: Fen | 011 |
| Carr | 012 |
| Bog | 013 |
| Salt marsh | 014 |
| Cave/Well/Tunnel: Threshold | 021 |
| Dark zone | 022 |
| Building: Inside | 031 |
| Outside | 032 |
| Garden: Domestic | 041 |
| Waste ground: <25% veg. cover | 051 |
| >25% veg. cover | 052 |
| Arable: Cereal crops | 061 |
| Root crops | 062 |
| Fodder crops | 063 |
| Grass ley | 064 |
| Market garden/allotment | 065 |
| Grassland: Ungrazed | 071 |
| Lightly grazed | 072 |
| Heavily grazed | 073 |
| Mown | 074 |
| Scrubland: Dense | 081 |
| Open with herbs/grass | 082 |
| Woodland: Dense | 091 |
| Open with scrub | 092 |
| Open with herbs/grass | 093 |
| Acid heath/moor: Moss/lichen | 101 |
| Grass/sedge/rush | 102 |
| Heather | 103 |
| *Vaccinium* (bilberry) | 104 |
| Mixed | 105 |
| Sand dune: Bare sand | 201 |
| Tussocky | 202 |
| Dense sward | 203 |
| Dune slack | 204 |
| Dune heath | 205 |
| Other: If none of above | 301 |

**D** 2nd order habitats;
1 tick (where applicable):

| | 16, 17 |
|---|---|
| Cold frame | 01 |
| Rockery | 02 |
| Flower bed | 03 |
| Lawn | 04 |
| Compost/refuse heap | 05 |
| Dung heap | 11 |
| Hay (or other) stack | 12 |
| Potato (or other) clamp | 13 |
| Hedge | 21 |
| Roadside verge | 22 |
| Embankment/cutting | 23 |
| Woodland ride/firebreak | 24 |
| Wood fence | 25 |
| Dry stone wall | 31 |
| Wall with mortar | 32 |
| Quarry face | 33 |
| Quarry floor | 34 |
| Natural cliff face | 35 |
| Rock pavement | 36 |
| Stabilised scree | 37 |
| Unstabilised scree | 38 |
| Grike | 41 |
| Road/path | 51 |
| Dry water course bed | 61 |
| 'Dry' ditch | 71 |
| Wet ditch | 72 |
| Shore/water edge/strandline | 81 |
| Vegetated stream | 91 |
| Unvegetated stream | 92 |
| Puddle | 93 |
| Pond <1 acre (0.4 hectare) | 94 |
| Flood patch | 95 |

Biological Records Centre June 1970    RA 15

140

| E MICROSITE (animal actually found under, on or in); 1 tick (obligatory): 36, 37 | | |
|---|---|---|
| Stones | 01 | |
| Shingle | 02 | |
| Soil/sand | 03 | |
| Litter | 11 | |
| Tussocks | 21 | |
| Bark (living trees or shrubs) | 31 | |
| Dead wood | 32 | |
| Dung | 33 | |
| Carrion | 34 | |
| Bracket fungi | 35 | |
| Ant colony (specify if possible) | 41 | |
| Bird/mammal nest (specify below) | 51 | |
| Rock | 61 | |
| Stone or brick work | 62 | |
| Shore line jetsam | 71 | |
| Human rubbish/garbage | 81 | |
| Other (specify below if possible) | 91 | |

| F HABITAT QUALIFIERS; 1 tick in each section, where applicable: | 38 | |
|---|---|---|
| (a) Building: Cellar | 1 | |
| Inhabited/public | 2 | |
| Uninhabited/outbuilding | 3 | |
| Ruin | 4 | |
| Greenhouse (heated) | 5 | |
| Greenhouse (unheated) | 6 | |

| | 39 | |
|---|---|---|
| (b) Shore: Intertidal | 1 | |
| Splash zone | 2 | |
| Between splash zone and 100m | 3 | |
| 100–1000m above H.W.M. | 4 | |

| | 40 | |
|---|---|---|
| (c) Encrustations: Moss | 1 | |
| Lichen | 2 | |
| Pleurococcoids | 3 | |

| | 41 | |
|---|---|---|
| (d) Waterspeed: Fast | 1 | |
| Slow | 2 | |
| Standing | 3 | |

| | 42 | |
|---|---|---|
| (e) Watercourse bed: Rocks | 1 | |
| Pebbles | 2 | |
| Sand | 3 | |
| Silt | 4 | |
| Peat | 5 | |

| G LIGHT LEVEL; 1 tick (obligatory): | 43 | |
|---|---|---|
| Full daylight | 1 | |
| Half-light/dusk/dawn | 2 | |
| Dark | 3 | |

**Other information e.g.**
Abundance,................(number)
collected per................(please state state unit of time/
space/volume) (53)   Aspect/degree of slope (54)
Behaviour (55)   Food (56)   Predators and Parasites (57)
Age structure and Sex ratio (58)   etc. (59)

| H SOIL/LITTER DETAILS, terrestrial habitats only; 1 tick in each section where applicable: | 44, 45 | |
|---|---|---|
| (a) Litter mainly: Oak | 01 | |
| Beech | 02 | |
| Birch | 03 | |
| Sycamore | 04 | |
| Mixed deciduous | 05 | |
| Coniferous | 11 | |
| Mixed decid./conif. | 21 | |
| Gorse | 31 | |
| Hawthorn | 32 | |
| Heathers | 33 | |
| Sea Buckthorn | 34 | |
| Litter/veg. mainly: *Carex* | 41 | |
| *Molinia* | 42 | |
| *Dactylis* | 43 | |
| *Festuca* | 44 | |
| *Bromus* | 45 | |
| *Brachypodium* | 46 | |
| Grass—species unknown | 47 | |
| Mixed grass/herbs | 51 | |
| Nettles | 61 | |
| Reeds (*Phragmites*) | 62 | |
| *Juncus* | 63 | |
| Bracken | 64 | |
| Other (specify below) | 71 | |

| | 46 | |
|---|---|---|
| (b) Litter age: Fresh | 1 | |
| Old | 2 | |
| Both | 3 | |

| | 47 | |
|---|---|---|
| (c) Litter cover: Exposed | 1 | |
| Protected by thin veg. | 2 | |
| Protected by thick veg. | 3 | |

| | 48 | |
|---|---|---|
| (d) Soil/exposed rock: Calcareous | 1 | |
| Non-calcareous | 2 | |

| | 49 | |
|---|---|---|
| (e) Soil: Heavy clay | 1 | |
| Clayey | 2 | |
| Peat | 3 | |
| Loam | 4 | |
| Sandy | 5 | |
| Pure sand | 6 | |

| | 50 | |
|---|---|---|
| (f) Humus type: *Mull* | 1 | |
| *Mor* | 2 | |

| I LOCATION OF ANIMAL; 1 tick in each section where applicable: | 51 | |
|---|---|---|
| (a) Horizon: >3m above ground | 1 | |
| <3m above ground | 2 | |
| On ground surface | 3 | |
| In litter | 4 | |
| <10cm in soil | 5 | |
| >10cm in soil | 6 | |

| | 52 | |
|---|---|---|
| (b) Position: In open | 1 | |
| In crevice | 2 | |

Isopod habitat card – *front* (p. 140) and *reverse*

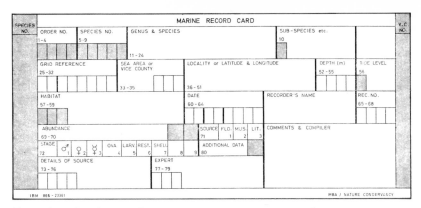

General marine data card

marine invertebrates, whilst the card illustrated (front and reverse side) on pp 140–1 is used to record the distribution and habitat preferences of one particular group of animals – the terrestrial isopods or woodlice. Data from cards of this sort are eventually stored on magnetic tape or disc, and can be selectively retrieved in cartographic or tabular form to meet the needs of ecologists and organisations concerned with conservation.

**Labelling procedure and materials**

A good policy is to label the material in the field at the time of collection rather than to commit information to memory with a view to making notes at a later date. Similarly, the replacement of temporary labels with later permanent ones is a source of potential error.

Labels should, as far as possible, be placed inside the receptacles containing the specimens and not gummed or tied to the outside. Only the finest possible paper should be used for labels, and ordinary pulp paper is quite useless. Labels for dry-preserved material should be made from a good quality rag paper, and for those that are to be immersed in preservative solutions a goat-skin parchment paper is recommended.

A good quality waterproof indian ink is suitable for writing both dry labels and those that are to be immersed, but before immersing labels, the ink must be allowed to dry thoroughly. As a further check against the possibility of the ink running, many experienced collectors advocate washing the label in water and allowing it to dry again before immersing it in the preservative. Labels for immersion can also be written with a soft lead pencil, but ball point pens should *never* be used. Labels can be most thoroughly protected by dipping them quickly into a bath of paraffin wax after they have been written.

142

**Packing**

Although with plastic tubes and bottles the risk of breakage in transit is virtually negligible, material should be carefully packed as the specimens themselves can easily suffer damage from vibration and shaking. Tubes or bottles, suitably sealed or separately wrapped in paper, should be packed in a wooden box or tin, lined and well padded with fine shavings, wood or cotton wool, or soft crumpled paper.

Delicate dry specimens, such as certain corals, should be packed well padded in an inner container which in turn should be packed and padded in an outer container. Specimens of this sort should never be fastened to the outer container, because any mechanical shock to the latter will be transmitted to the specimen. Even sawdust may transmit vibration if it has become strongly compacted.

Fluid-preserved specimens are much less likely to be damaged by shaking during transport if the tubes containing them are almost full of fluid. Alcohol, however, expands with heat and there is a danger of breakage unless a small air space is left in the tubes. With spirit-preserved material evaporation may also present a problem, especially in hot climates. This can be overcome by using self-sealing plastic tubes, or if cork tubes are used by dipping the corked ends two or three times in melted paraffin wax, allowing each coat of wax to set before the next is applied.

When a considerable number of tubes containing liquid-preserved material have to be packed, a better method than dry packing is to place them in a large jar of the same preservative as that in the tubes. For this, a layer of cotton-wool, at least 1 cm or so thick, is placed in the bottom of the jar and saturated with preservative solution. The separate tubes are filled completely with preservative and plugged with saturated cotton-wool. The plugs should just overhang the edges of the tubes and should be covered on the inner side with tissue paper to prevent the specimens from becoming entangled with wool fibres. The tubes are then placed upside down in the jar and covered with preservative solution. Any spaces between the tubes should be packed with sufficient cotton-wool to prevent rattling. Another thick layer of wool is then placed on top of the tubes and the jar filled almost to the top with preservative, leaving a small air space for expansion. If necessary, several layers of tubes can be packed in the same jar. Finally, the lid of the jar is secured in place making sure that the sealing washer is properly seated.

Larger specimens that are unsuitable for tubes may sometimes have to be packed together in tanks or large jars of liquid. Each specimen together with its label should be wrapped in muslin. When this method is used the jar or tank need not be full of liquid provided that it is well padded with cotton-wool or other material soaked in the liquid. When unwrapped specimens are packed, however, the jars should be well filled.

### Dispatch of specimens from overseas

The transport of significant quantities of spirit, and therefore of specimens preserved in it, is subject to various formalities and restrictions. For example, packages containing it may have to be treated as deck cargo on board ship. Collectors are advised to enquire about such regulations before dispatching material.

The British Museum (Natural History) can only undertake to pay the cost of shipping material to London if such an arrangement is agreed in advance. In such cases large items may be handed to a shipping agent who will then contact the Museum to receive a guarantee that charges will be paid on receipt. All packages, whether being sent by mail or by freight, should be addressed to The Director, British Museum (Natural History), Cromwell Road, London SW7 5BD. By special arrangement with H.M. Customs, packages so addressed will normally be sent under seal direct to the Museum. Suitably printed address labels will be supplied to prospective collectors on request. As soon as a consignment has been dispatched, the Museum should be notified by airmail letter, stating the date of dispatch and the method of delivery (eg. by surface mail or freight). In the case of material sent by freight the name of the shipping agent should also be given.

# Index